A Field Guide
to the Trees
and Shrubs
of the Southern
Appalachians

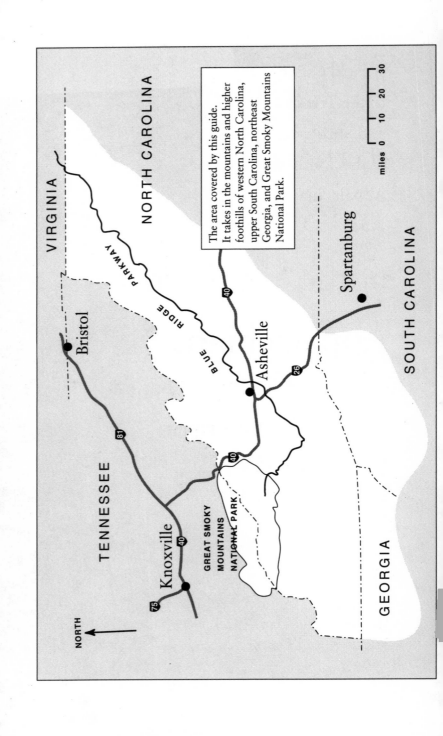

The area covered by this guide. It takes in the mountains and higher foothills of western North Carolina, upper South Carolina, northeast Georgia, and Great Smoky Mountains National Park.

A Field Guide
to the Trees
and Shrubs
of the Southern
Appalachians

Robert E. Swanson

Illustrations by
Frances R. Swanson

The Johns Hopkins University Press
Baltimore and London

© 1994 The Johns Hopkins University Press
All rights reserved.
Printed in the United States of America on acid-free paper
9 8 7 6 5 4 3 2

The Johns Hopkins University Press
2715 North Charles Street
Baltimore, Maryland 21218-4363
www.press.jhu.edu

Library of Congress Cataloging-in-Publication Data

Swanson, Robert E., 1915–
    A field guide to the trees and shrubs of the southern Appalachians /
Robert E. Swanson ; illustrations by Frances R. Swanson.
        p.   cm.
    Includes bibliographical references and index.
    ISBN 0-8018-4555-6 (alk. paper).—ISBN 0-8018-4556-4 (pbk. : alk.
paper)
    1. Trees—Appalachian Region, Southern—Identification.   2. Shrubs—
Appalachian Region, Southern—Identification.   I. Title.
QK122.3.S88   1993
582.160975—dc20   92-41572

A catalog record for this book is available from the British Library.

To the memory of my mother and father

Contents

Preface

The keys, descriptions, and illustrations in this field guide are based mainly on the more than five thousand plant specimens I collected between 1980 and 1991. During this twelve year period, often accompanied by my wife, I made more than three hundred excursions into the forested and mountainous areas of eleven states. An essential part of each excursion was a hike on mountain trails. On these plant-hunting trips, I walked more than 5,400 miles.

All the keys, descriptions, and illustrations are new, but I freely consulted the work of many authorities. To paraphrase the words of Sir Isaac Newton, we all stand on the shoulders of giants.

One of my personal giants was the late Helen M. Gilkey, professor of botany, Oregon State University. Not only did Dr. Gilkey instruct me in plant taxonomy, but she encouraged me to become a self-taught naturalist, constantly reminding me that biological expertise arises from sharp eyes and eagerness to understand, and that the better I know something, the more deeply I can explore it. Through her teaching and affection, Dr. Gilkey has given me lifelong inspiration.

I also wish to acknowledge the valuable aid and suggestions given by others, especially—

The late Arthur H. Graves, curator emeritus, Brooklyn Botanic Garden, who graciously replied to all the letters I wrote him about various aspects of plant taxonomy.

Elbert L. Little, Jr., chief dendrologist (retired), U.S. Forest Service, with whom I exchanged many letters and telephone calls about plants in general and trees in particular.

James Perry, professor of biology, University of North Carolina at Asheville, who allowed me to accompany him and his students on numerous plant-collecting trips into the Carolina mountains.

Donald J. Leopold, associate professor, College of Environmental Science and Forestry, State University of New York, who read my manuscript with care and offered many valuable suggestions.

Tom G. Sicama, director of field studies, School of Forestry and Environmental Studies, Yale University, whose helpful criticism of the major keys I appreciate.

Special thanks go to the staff of the Johns Hopkins University Press for helpful suggestions, and especially to Richard T. O'Grady, science editor, for his guidance and patience. Without his assistance the publication of this work would have been impossible.

And finally, to my wife, Frances, I express sincere thanks not only for her fine work on the illustrations but for her constant help throughout in checking the manuscript and for the many excellent suggestions that have so greatly improved the final product.

A Field Guide
to the Trees
and Shrubs
of the Southern
Appalachians

Introduction and
Terminology

The southern Appalachians consist of the Great Smoky Mountains and the southern Blue Ridge, together with such lesser ranges as the Craggy Mountains, the Black Mountains, and the Snowbird, Nantahala, and Balsam. They make up the mountainous areas of North Carolina, South Carolina, Georgia, and Tennessee. This is a world where you see mountains rolling like waves, a high, tumbled world fading into the haze, where as many as 130 different species of flowering trees and 11 conifers can be found—more total species than grow in all of Europe.

There are boreal forests at the highest elevations, mainly above six thousand feet, where the dominant trees are Fraser fir and red spruce. The higher ridges and mountaintops also bear treeless areas called grass and heath balds, the latter with a shrubby cover of mainly Catawba rhododendron, mountain-laurel, and various species of blueberry.

Northern hardwood forests are on cool, well-drained mountainsides; oak-hickory forests on warm, open slopes and ridgetops, beech-maple forests in cool, moist places; and pine-oak and northern riverine forests along the banks of larger creeks and rivers.

Richest of all are the cove hardwood forests, growing in protected coves and hollows at lower and middle elevations, where soils are deep, fertile, and moist. Here, in naturally terraced valleys, as many as thirty different trees may be found: yellow-poplar, sugar maple, American beech, Carolina silverbell, basswood, red maple, yellow birch, sweet birch, and yellow buckeye are among the taller trees; redbud, sweetleaf, sourwood, flowering dogwood, and Fraser magnolia are some of those in the subcanopy.

And beneath these trees is a glorious wealth of native shrubs and woody vines: rhododendron, azalea, minnie-bush, mountain-laurel, dog-hobble, bearberry, buffalo nut, strawberry-bush, mountain winterberry, withe-rod, fetter-bush, Dutchman's-pipe, greenbrier, moonseed, grape, Virginia creeper.

Added to the many native trees and shrubs are plants that humans have brought from other parts of the world, deliberately or by accident. With these exotics supplementing the original population of species, the area offers one of the richest and most diversified temperate-zone floras in the world. Some of the species have become so thoroughly naturalized that you can no longer assume an unknown plant is native simply because you find it growing wild.

This book is for everybody who wants to learn to recognize the trees, shrubs, and woody vines that one is likely to find in the mountains and the adjoining higher foothills of the southern Appalachians. It covers all the common native species, and many of the rare ones, that grow not only in the forests but in high mountain meadows, on heath balds, in long-abandoned fields, and along fencerows and roadsides. The keys and descriptions present characters of twigs and winter buds, as well as those of leaves, that can be found at any time of year.

Several species that are included are native to other parts of the United States or to other countries but have become thoroughly established in our area. Examples are multiflora rose (*Rosa multiflora*), Japanese honeysuckle (*Lonicera japonica*), and ailanthus (*Ailanthus altissima*). Also considered are species that were brought here for their utility or beauty and have escaped from cultivation. Familiar examples are English ivy (*Hedera helix*), rose-of-Sharon (*Hibiscus syriacus*), and apple (*Malus sylvestris*). A few plants, such as common lilac (*Syringa vulgaris*), are included because they often persist around abandoned farmsteads and in other places long after cultivation has ceased.

Let me begin with some definitions.

Trees, shrubs, and vines. Trees are considered here to be woody plants that have a well-defined trunk at least two centimeters in diameter and that usually reach a height of four meters or more. Although the difference between trees and shrubs is not always clear, shrubs typically are smaller and usually have several stems growing from the ground instead of a single trunk. A simpler definition: if you can walk under it, it is a tree; if you have to walk around it, it is a shrub. In most cases, however, it is neither practical nor necessary to separate trees from shrubs, and the keys in this book do not do so.

A vine is a climbing, sprawling, or trailing plant whose stem requires external support. Vines are usually considered shrubs, and where a support is lacking they often grow as conventional shrubs, as poison-ivy does.

The book includes several plants that are on the border between woody and herbaceous. One can argue that these borderline plants do not belong here, but I have added them as a convenience. These plants are pipsissewa (*Chimaphila maculata*), leather-flower (*Clematis viorna*), virgin's bower (*Clematis virginiana*), trailing-arbutus (*Epigaea repens*), aromatic wintergreen (*Gaultheria procumbens*), partridge-berry (*Mitchella repens*), and periwinkle (*Vinca minor*).

Deciduous and evergreen plants. The word *deciduous* comes from the Latin *decidere,* "to fall off." Deciduous woody plants shed their leaves in autumn. Evergreens keep their leaves over winter, and the leaves stay green. If a plant is evergreen, leaves are present on twigs of the previous year as well as on twigs of the current year. The conifers in our area retain their green leaves through winter, but they are not the only evergreen plants. Some broad-leaved trees and shrubs, such as mountain-laurel (*Kalmia latifolia*) and American holly (*Ilex opaca*), retain their leaves throughout the year. Half-evergreen plants keep their green or bronzy leaves in mild winters and in protected locations. Examples are the privets (*Ligustrum*) and the greenbriers (*Smilax*).

The Structure of Woody Plants

Parts of a leaf. During the growing season, a woody plant is most readily identified by its leaves. A leaf consists of a *blade,* usually flat and green, and a *petiole,* or leafstalk. A leaf with only one blade, as in the willows and honeysuckles, is called a *simple* leaf. One with several blades, or *leaflets,* is a *compound* leaf. Familiar examples of compound leaves are those of hickories, roses, and black walnut.

Compound leaves having leaflets that radiate, like the spokes of a wheel or fingers of a hand, from the top of the petiole are called *palmately* compound, as in yellow buckeye and Virginia creeper. Those having leaflets attached along the sides of the petiole are *pinnately* compound, as in walnut and the sumacs. If the petiole itself is branched, the leaf is *twice compound,* as in devils-walkingstick and Kentucky coffeetree.

The petiole supports the blade and also attaches the leaf to the twig. The place where a leaf is attached, or where a leaf scar occurs on a winter twig, is a *node.* At the base of the petiole on some leaves, one on each side, are two small scaly or leaflike organs called *stipules.* A few plants have stipules that are conspicuous or persistent, and these are mentioned in the descriptions. Many plants do not have stipules.

Arrangement of leaves. Careful observation of normal twigs will show that leaves are usually arranged in one of three ways. Those attached to the twig in pairs at the same level are said to be *opposite,* leaves attached singly on the stem are *alternate,* and three or more leaves borne at the same node are *whorled.*

Venation. Every leaf has *veins,* the vascular bundles that enter the blade from the petiole. The arrangement of the veins, their prominence and manner of branching, are useful characters in identification. If several important veins arise at or near the base of the blade, the leaf is *palmately veined.* If a blade has one large vein, or *midrib,* and several to many smaller veins branching from it, the leaf is *pinnately veined.* In

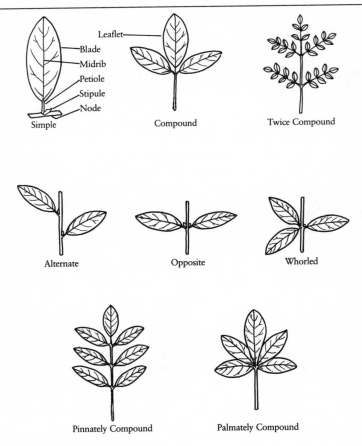

Simple

Compound

Twice Compound

Alternate

Opposite

Whorled

Pinnately Compound

Palmately Compound

FIGURE 1   Parts of a leaf; types and arrangements of leaves.

a few woody plants, such as the dogwoods, the veins run nearly parallel to the margin, the upper veins ending at the tip of the leaf. This is a modified form of pinnate venation.

Shape of leaves. Terms used for describing leaf shape are illustrated in figure 2 and defined in the glossary. The shapes of leaves can vary in a vast number of ways, but we have only a few words to describe them. For this reason we designate intermediate forms by combining two terms; that is, ovate-lanceolate is narrower than typical ovate, and so

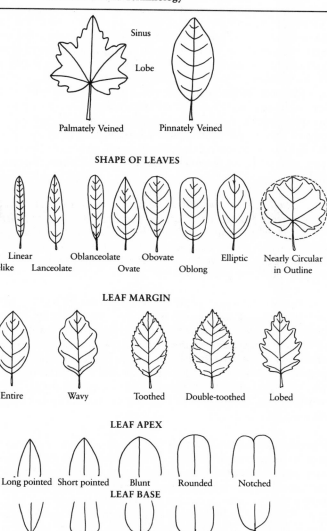

FIGURE 2   Leaf shapes, bases, apexes, and margins.

on. In addition, an adverb frequently modifies a term, as in broadly ovate or narrowly elliptical. Words describing shape refer only to the general outline of the blade. They do not indicate the size of the leaf or the pattern of the veins, nor do they describe the apex, base, or margin of the leaf.

In observing the shape of leaves on a plant, take care to select normal or typical leaves. As a rule, leaves on seedlings, sucker growth, or vigorous sprouts are not typical; they are often much larger and may differ greatly in shape and other characteristics. The keys and descriptions in this book are based on average leaves—generally, those on flower-bearing or fruit-bearing twigs—and on mature leaves, not early spring growth.

The apex and base of the leaf. The tip, or *apex,* of the leaf is described in simple, everyday words, the most common being *long pointed, short pointed, blunt, rounded,* and *notched.* The base also is described in common words: *wedge shaped, rounded, straight across, heart shaped,* and *unequal or unsymmetrical.* All these words are illustrated in figure 2 and defined in the glossary.

The margin of the leaf. A leaf having a smooth margin, without teeth or indentations of any kind, is called *entire.* Many leaves have a *toothed* margin, with the teeth pointing outward, pointing forward (toward the leaf apex), or curving inward (toward the leaf). The leaves of some woody plants are *lobed;* that is, the blade is divided into lobes by shallow to deep notches, or *sinuses,* as in red oak and often in sassafras. Other special terms referring to the leaf margin, though of interest to botanists and to students with a technical interest in woody plants, are not used in this book.

The surface of leaves. Many leaves are smooth and green, but some have a bloom, and others are covered with hairs of various kinds. Leaves (and stems also) are *glabrous* if they are smooth and without hairs. A smooth leaf is *glaucous* if it is covered with a whitish or bluish *bloom,* a sort of fine waxy powder, usually easy to rub off. A surface coated with small scalelike or branlike particles is *scurfy.* Among terms used to describe hairiness are *woolly, silky, velvety,* or simply *hairy.* If the hairs are stiff, they are called *bristly.*

Other features of the leaf surface are glandular or resinous dots or droplets, gland-tipped hairs, and variously clustered or branched hairs. Some leaves have internal glands, which show up as translucent dots when the leaf is held to the light.

It is also well to note leaf texture: some leaves are thick and leathery, whereas others are thin, rather soft, and often somewhat translucent.

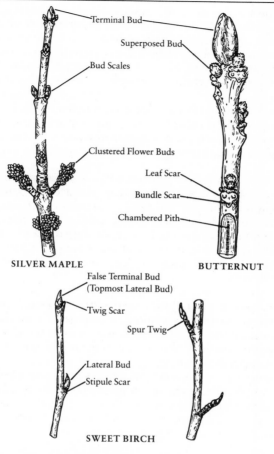

FIGURE 3   Winter twigs showing buds, leaf scars, and other winter characters.

Some of the terms used here are common and need no explanation; the rest are defined in the glossary.

Twigs. A twig is a small branch and usually constitutes the growth of several years. The term *young twig,* as used here, means a young woody stem, the terminal part of a branch, one year old or less. An *older twig* is the part that is older than one year.

Leaf scars. A leaf scar is the mark left on a twig when a leaf falls. It represents the point of attachment between petiole and twig. These scars take on a wide variety of sizes, outlines, and patterns, and all vary according to the species of plant. They often appear as circles, crescents, shields, hearts, and even horseshoes—all of different dimensions.

The arrangement of leaf scars is the same as that of leaves: alternate, or scattered singly along the twig; opposite, or placed in opposing pairs on the twig; and whorled, where three or more scars are borne in a circle.

A word of caution: to examine the arrangement of leaf scars on a plant to be identified, select an average twig for study. Avoid short twigs on which the scars are densely crowded, since the arrangement may be difficult to determine.

Bundle scars. Each leaf scar bears on its surface one or more small dots, or *bundle scars,* which represent the vascular bundles that extended from the twig into the petiole before the leaf was shed in autumn. Their number and shape are usually constant for each species. The most common numbers of bundle scars are one and three. The dots on some leaf scars, such as those of black tupelo, are relatively large and easy to see. Others, much smaller or less distinct, call for the use of a magnifying lens.

Buds. The *buds* of most woody plants are clearly visible for eight or nine months of the year, dispelling the popular notion that buds suddenly pop out in the spring. After reaching their full development by August, buds normally go through autumn and winter without change, resuming growth in the spring.

A bud contains the growing point of a stem, temporarily at rest because of an unfavorable change in growing conditions. Borne on this tiny stem are undeveloped leaves or flowers, or both, folded tightly together and usually protected by one or more covering organs known as *bud scales.*

Bud scales. For each species or group of species the number of bud scales is usually constant. Those of the willows and a few other plants are *solitary,* or single, each scale sheathing the inner parts of the bud like a hood. The *valvate* buds of yellow-poplar and the dogwoods, among others, have two outer scales meeting at the edges, somewhat in the manner of clamshells cupped together. The most common arrangement, however, is a number of *overlapping* scales like shingles on a roof. Finally, a few plants, such as witch-hazel and Carolina buckthorn, develop *naked* buds, which have no scaly covering but are protected by the outermost pair of rudimentary leaves.

Leaf buds and flower buds. If two sizes of buds appear on the same plant—as they often do on twigs of red maple, sweetleaf, redbud, huckleberries, and several other plants—the smaller buds contain tiny leaves and are called *leaf buds.* The term is something of a misnomer, however, because the bud produces not only leaves but a young stem similar to the one that bore it. The larger *flower buds* contain undeveloped flowers. Some plants, such as the basswoods, serviceberries, and sycamore, have buds that contain both flowers and leaves.

Terminal buds and lateral buds. Buds are classified, among other ways, by their location on the twig. A *terminal* bud is at the exact tip of the twig, and *lateral* or *axillary* buds develop in the upper angle where a leaf joins a twig, directly at or above a leaf scar. The twigs of some woody plants do not have a terminal bud; the uppermost lateral bud assumes the end position. In the next growing season, the twig is lengthened by the growth of this bud.

In placing a plant in its proper genus or species, it is often important to know whether the twigs have a terminal bud. A close examination of the twig tip will tell if a terminal bud is present. If it is, the bud usually stands parallel to the twig, pointing straight ahead, and often is larger than the lateral buds. If a lateral bud has taken the end position, it usually appears tipped, not pointing straight ahead; it is scarcely larger than the other buds; and on the side of the twig away from the topmost leaf or leaf scar, there is a short scar or stub showing the dead end of the season's growth. This scar is usually much smaller than the leaf scars and has a different appearance.

Thorns and prickles. A thorn, as its name suggests, is a stiff, woody, sharp-pointed projection. In this book the term is used to mean a modified twig, leaf, or stipule. If a thorn is a modified twig, it occurs in the axil of a leaf or terminates a branch. Sometimes a thorn bears leaves and buds—further evidence that it is a modified twig. A thorn that is a modified leaf is attached just below a lateral bud. Thorns that are modified stipules are borne in pairs at the base of a leaf petiole or at the sides of a leaf scar.

A *prickle* is a slender, sharp-pointed outgrowth from the young bark or epidermis of a twig. Prickles are not connected to the vascular system of the plant and, as a rule, are easily broken off.

Pith. The center or core of a twig is called the *pith.* Because it is usually lighter or darker than the wood surrounding it, pith is usually readily distinguished and can be revealed from two different angles, depending on the way the twig is cut or sliced. If a twig is carefully cut crosswise, the pith can be seen in cross section. If it is sliced lengthwise, the pith is seen in longitudinal section.

The twigs of most trees and shrubs have a solid pith—the core is continuous and uniform. In some species, however, the pith is *chambered*. When the twig is split or sliced lengthwise, the partitions forming the walls between the chambers look like the rungs of a ladder. The chambers may be hollow, as in walnut and hackberry, or filled with loose tissue, as in yellow-poplar, pawpaw, and black tupelo. In most species of grape, the pith has a single woody partition at each node.

The shape of the pith in cross section not only is interesting to explore but also may help in tracking down the genus or species. When making a cross-sectional cut, choose a place between the nodes and cut carefully, using a sharp knife or other instrument. Twigs reveal pith of varying shapes—circular (most species), triangular (alders), five-pointed (oaks), or sometimes of no particular shape—depending on the kind of woody plant.

The color of the pith may also be important. In most species the pith of older twigs is white, but in Kentucky coffeetree it is salmon red, and in ailanthus and the sumacs it is yellowish brown or brown.

Bark. Bark is one of the most important features in identifying woody plants. Because it is difficult to describe, however, its characteristics are best learned by observation. Among the bark features that are especially distinctive are the peeling and shredding of river birch, ninebark, and gooseberry, the mottled patchwork of sycamore, the silvery appearance of autumn-olive, and the white strips of Carolina silverbell and striped maple. In some species, such as blackhaw and flowering dogwood, the trunk bark is divided into small, thick, squarish scales that resemble alligator hide. In others, such as sweet birch and the cherries, it is smooth and strongly marked with long horizontal lines.

In many species the bark of twigs emits a characteristic odor—pleasantly fragrant or spicy, as in sweetfern, spicebush, sweetshrub, yellow and sweet birches, magnolia, and yellow-poplar, or unpleasant or foul, as in chokecherry, skunk currant, and ailanthus. Bruised bark of black cherry and the domestic cherries has an odor like crushed cherry pits or bitter almonds. Members of the willow family—cottonwood, poplar, willow—have bark that is exceptionally bitter.

Anyone interested in identifying woody plants should become familiar with bark features, since they are useful throughout the year.

Other features. A short, stubby, slow-growing twig that bears one or more leaves or buds, or both, and is roughened by crowded leaf scars is called a *spur,* or spur twig. A few short twigs may be found on almost any tree or shrub, but spurs are especially common on such plants as pin cherry, apple, the domestic cherries, birches, currants, and gooseberries.

A corky spot or area in the bark, serving to admit moisture and air to the living tissues beneath when the bark was first formed, is known as a *lenticel*. Lenticels are features of many smooth-barked woody plants, and elongated lenticels are particularly useful in identifying the birches and cherries.

Fruits

In popular usage "fruit" usually means a structure with juicy, edible flesh, such as apple, cherry, pear, blueberry, or huckleberry. Many persons do not think of pinecones, acorns, hickory nuts, locust pods, and maple keys as fruits. Nevertheless, all are fruits in the botanical sense. *Fruit* as used in this book refers to any seed-bearing organ of a tree, shrub, or woody vine.

The fruit of the coniferous trees in our area is a scaly cone (the cone is berrylike in eastern redcedar). The seeds are borne at the base of the cone scales, and when the scales open the seeds are seen to be *naked* (not enclosed by an ovary). Such plants are called naked seeded, and botanically they are referred to as *gymnosperms*.

All of our broad-leaved trees, as well as our shrubs and woody vines, have their seeds enclosed within ovaries and are called *angiosperms*. The fruit of these plants is the ripened product of a flower, together with any appendages and withered flower parts that may be attached.

Fruits of angiosperms may be classified in various ways. Botanists and horticulturists often distinguish them by origin: they trace the fruits back to the type of flowers that produced them. In this book, however, it seems most useful to classify fruits by whether they are dry or fleshy at maturity. Though the popular notion of fruits includes only the fleshy ones, many woody plants produce various types that are dry when they mature.

Some of the more common fruits of trees and shrubs are described in the following list:

Some Dry Fruits That Split Open When Ripe
True pod splitting open on opposite sides (*legume*): indigobush, honey-locust, redbud, black locust.
Podlike fruit splitting along one side only (*follicle*): spiraea, ninebark; a magnolia "cone" is a compact group of follicles.
Podlike fruit with usually more than one cell (*capsule*); splits lengthwise when ripe: rhododendron, mountain-laurel, sourwood.

Dry Fruits That Do Not Split Open When Ripe

Small dry fruit, unwinged but often tipped with hairs or bristles (*achene*): a fruit head of sycamore consists of numerous achenes packed tightly together.

Small fruit equipped with a wing (*samara*): yellow-poplar, maple, ash, elm.

Achenelike fruit with a relatively thick wall (*nutlet*): American hornbeam, basswood, hophornbeam.

Fruit with a hard outer shell and a kernel (*nut*): chestnut, hickory, oak (acorn).

Common Fleshy Fruits

Fruit soft and fleshy throughout, usually several seeded but sometimes with only one seed (*berry*): persimmon, grape, greenbrier, currant (but not mulberry or blackberry, whose fruit is a cluster of small structures called *drupelets*).

Fruit resembling a berry but with a core like that of an apple (*pome*): serviceberry, mountain-ash, hawthorn, chokeberry.

Stone fruit with the outer part fleshy, the inner part bony (*drupe*): poison-ivy, viburnum, peach, hackberry, dogwood, sassafras. In Osage-orange, many drupes are compacted in a grapefruit-sized ball.

Some woody plants retain their fruits through winter. These are most likely to be dry fruits, since the fleshy types usually fall to the ground soon after they ripen or are eaten by wildlife or sometimes by humans. The fleshy fruits that do persist are likely to be dried and shrunken, such as the pomes of the serviceberries and hawthorns, or those with thin flesh, such as the drupes of the hackberries.

In some species, such as ashes and hollies, the male (*staminate*) and female (*pistillate*) flowers are borne separately on different plants, and consequently not all plants bear fruit. Many species of trees and shrubs do not produce fruit every year, and most trees do not bear until they reach a certain age. In spite of these limitations, however, fruits are useful in identification whenever they can be found.

Poisonous Plants

To those who explore the mountains on foot, poison-ivy is public enemy number one. All parts of the plant contain resinous oils that irritate and blister the skin of sensitive individuals. Hence its leaves, stems, flowers, and fruit cannot be handled safely by most persons.

Typically, poison-ivy (*Toxicodendron radicans*) is a vine that climbs, by means of aerial rootlets, on the trunks of trees, over fences and hedges, along roadsides, and even in gardens and lawns. Individual plants climbing trees for support often send out twigs horizontally in such a manner that they may be mistaken for part of the tree itself. The aerial roots of such plants give the stem a fuzzy appearance that helps one to recognize them. If a support is lacking the plant spreads by running roots in areas of partial shade, and it may even take the form of a single-stemmed shrub.

Poison-ivy has these characters: alternate, compound leaves with three leaflets ("Leaflets three, better flee!"), the central leaflet long stalked; stems usually vinelike, dotted with lenticels and bearing aerial rootlets; hairy, naked terminal buds; and berrylike fruits, white or cream colored, borne in drooping clusters. An illustration of poison-ivy can be found on page 253.

Poison-sumac (*Toxicodendron vernix*) is much less common in the mountains and is confined to wet or marshy ground. Any shrub that has pinnately compound leaves and is growing wild in a wet or swampy area should be viewed with suspicion until its identity is determined. Poison-sumac is described on page 253.

If a person who touches a poisonous plant is sensitive to the oils that produce dermatitis, a rash will erupt in twelve to forty-eight hours and persist for about two weeks. Antihistamines will reduce the itching, as will cool compresses of Burrow's solution. Hydrocortisone creams such as Cortaid and Lanacort are helpful and are available without a prescription. If the rash is too extensive to be easily treated, or if the itching is so severe it cannot be tolerated, consult a physician.

You can minimize the effects of the oils by scrubbing the skin vigorously with a detergent or strong soap, such as Octagon or Fels-Naptha, within thirty minutes of exposure. On clothing and camping equipment, the oils usually remain potent for a long time. These items should be washed in warm water, using a detergent that breaks down the oils.

Names of Plants

Every kind of woody plant in the southern Appalachians has a scientific name and one or more English or common names. The technical name, or Latin binomial, consists of two parts, the genus plus the species. The genus (generic) name is written first, with an initial capital, followed by the species (specific) name beginning with a small letter, both in italics. Thus we have a name that is intelligible to well-informed people in all countries. This remarkable system of nomenclature was devised by the great Swedish naturalist Linnaeus and is used in the naming and classification of every known tree, shrub, and woody vine, as well as all other living things.

To illustrate, let us consider two related trees that have many traits in common but show distinct differences that warrant placing them in separate species: *Quercus alba* (white oak) and *Quercus rubra* (northern red oak). Both belong to the same genus, *Quercus*—the Latin name of the oak. They differ in species: *alba* (white), and *rubra* (red).

A *species* is a group of organisms that have many distinctive features in common and reproduce their own kind. For example, if we plant the seeds of flowering dogwood and later observe that many of them have sprouted, we expect the seedlings to be flowering dogwoods like the parent trees.

Closely related species are grouped as a *genus* (plural, *genera*). We recognize these groups in our everyday language when we speak of the maples, the dogwoods, or the oaks, and we name the species when we say that a plant is a striped maple, a silky dogwood, or a black oak.

Related genera are combined to form a *family*. For instance, the apples, hawthorns, cherries, and plums, together with the shrubby genus *Rosa* (which gives the family its name) and numerous other plants, are members of the large rose family, Rosaceae. Families, in turn, are combined into larger groups, and so on.

For trees and treelike species, the scientific names and common names used in this book are taken from the *Checklist of United States Trees (Native and Naturalized)* by Elbert Little, Jr. Names of shrubs and woody vines follow, with a few exceptions, the *Manual of the Vascular Flora of the Carolinas,* by Radford, Ahles, and Bell.

Classification

Every field guide creates a kind of classification when it lists genera or species in identification keys according to certain distinguishing characteristics. The alphabetical listing of families, genera, and species in the index is another kind of classification. Usually, however, classification in biology refers to a "natural" grouping of organisms that shows their evolutionary relationships—the oldest families first and the most recently evolved last.

The arrangement of families in this book is that proposed by the German botanist Adolph Engler and is generally the same one followed in the *Manual of the Vascular Flora of the Carolinas* by Radford, Ahles, and Bell. In addition, evolutionary relationships are shown in the section "List of Trees and Shrubs Arranged according to Families," where species are assigned to genera, and genera to families, based on the system initiated by Engler, which has been the dominant scheme of plant classification for most of this century.

Habitat and Distribution

Habitat and distribution are based on information given in *Flora of the Carolinas* and in *Trees, Shrubs, and Woody Vines of Great Smoky Mountains National Park* by Arthur Stupka, combined with my own knowledge. Distribution is described by frequency of occurrence—ranging from common or fairly common to infrequent, occasional, and rare—and by states or areas of occurrence: Georgia (GA), North Carolina (NC), South Carolina (SC), and Great Smoky Mountains National Park (GSMNP). In some cases there is unavoidable overlap between NC and GSMNP, but NC generally refers to areas of North Carolina outside the park.

Measurements

Measurements in this guide are given in metric units, the standard for scientific communication. A millimeter (mm) equals about $\frac{1}{25}$ of an inch, a centimeter (cm) about $\frac{2}{5}$ of an inch, and a meter (m) 39.37 inches, or about 3.3 feet. A scale that converts millimeters and centimeters to inches is provided inside the cover of this book.

Illustrations

Most of our illustrations have been reduced to fit the page and hence do not show actual size, but some have been enlarged to show detail. Important measurements of plant parts are given in the descriptions of species; by referring to these measurements, the reader can readily estimate the amount of reduction or enlargement in the illustrations.

The Keys

How to Use the Keys

If you examine the leaves and twigs of a yellow-poplar tree in summer, looking closely at the external features, you will see that the leaves are alternate and simple, with a broad, slightly notched apex and four to six lobes; the twigs are encircled by a faint ring (stipule scar) at each node; and the terminal bud is flattened and blunt, with two valvate scales. Flowering dogwood has deciduous, opposite, simple leaves in which the veins curve strongly toward the leaf apex, nearly parallel to the margin, and the stalked flower buds are biscuit shaped. When characters such as these are arranged in an orderly manner, enabling us to compare an unknown specimen with a series of contrasting statements and thus arrive at the identity of a plant, we have a *key*.

Without a key you would have to compare your specimen with every description and illustration in the book—a time-consuming task. But valuable as they are, keys are not always easy to use. They require accurate observation and a correct decision at each step. A single mistake will lead you in the wrong direction and soon bring you to a dead end where none of the statements agree with your plant. If this happens, you must go back to the beginning and start over, carefully verifying each step.

It is of the utmost importance that you select typical leaves and twigs for examination. Avoid unusually large or small leaves, and in the leafless season, avoid twigs that appear extreme in any way. Especially do not choose stump sprouts, for these often bear leaves or leaf scars and buds that differ greatly from average growth.

In using the keys, you may note that a genus or species shows up in more than one place. This duplication is sometimes necessary because of variations within the group, or the second listing may compensate for common errors in identification.

Do not be discouraged if you fail to identify a plant correctly on the first trial. Even experts are not always certain. Keep trying, and you will soon be familiar enough with common species to recognize them at first sight. Others will take more experience and effort.

When you want to identify a woody plant from a specimen in hand, first go to the table of contents, which will help you to select the proper key.

The keys are constructed so that you are offered one pair of choices after another, each consisting of statements that contrast strongly with each other. Only one statement in each pair can be true—no fence straddling is possible!

The paired choices are prefaced by the same number (e.g., 1 or 1') and always begin with the same word. Numbering is consecutive, and each number is used twice in each key.

Your first choice will always be between number 1 and number 1'. For example, if you have consulted the table of contents and have determined that your specimen in hand is from a woody plant with alternate simple leaves, you will select key IV, beginning on page 32.

From key IV, the first point to be decided is whether your specimen is from a vine (no. 1) or from a plant that is not a vine (no. 1'). If the plant is a vine, you again must make a choice, this time between "Stems with tendrils" (no. 2) and "Stems without tendrils" (no. 2'), and so on.

Your final selection will lead to two names—one a common name, the other a Latin name. In some cases the Latin name is that of the genus only; in others, it is the species. If it is the name of a genus, turn to the page indicated, and another key will help you determine which species you have. All of the keys list the page where each genus or species is described.

You will need a good hand lens to use the keys effectively, particularly the winter keys. A lens of about 10× power will enable you to see many small features of buds and twigs, such as size and shape of leaf scars, number of bundle scars, the presence or absence of stipule scars, the color, size, and shape of buds, and the number and arrangement of bud scales.

Key I. Plants with Needlelike or Scalelike Leaves

1 Foliage leaves needlelike, more than 8 mm long:
  2 Needles in fascicles of two to five, each fascicle enclosed at base in papery
    sheath.. pine, *Pinus* (p. 95)
  2' Needles borne singly on twigs:
    3 Twigs roughened by peglike projections after needles fall; cones
      hanging down, with persistent scales;
      4 Needles blunt, flat, with short stalks .... hemlock, *Tsuga* (p. 98)
      4' Needles sharp pointed, four-sided (angles apparent when leaf is
        rolled between thumb and forefinger) .... red spruce, *Picea rubens*
                    (p. 98)
    3' Twigs with smooth, circular scars after needles fall; cones upright, with
      deciduous scales....................... Fraser fir, *Abies fraseri* (p. 100)
1' Foliage leaves scalelike, less than 5 mm long (or some leaves needlelike and
  as much as 12 mm long):
  5 Leafy twigs four-angled in cross section; leaves mostly scalelike on older
    trees, mostly needlelike on younger plants, but both types may be present;
    fruit fleshy, bluish, berrylike; inner wood of older trees red and aromatic
    .... eastern redcedar, *Juniperus virginiana* (p. 102)
  5' Leafy twigs much flattened, in sprays, often fanlike; all leaves scalelike, in
    alternate pairs, each pair at right angles to pairs above and below, closely
    overlapping; fruit a small, dry cone with leathery, blunt-pointed scales;
    inner wood white .... northern white-cedar, *Thuja occidentalis* (p. 102)

Summer Keys

Key II. Plants with Opposite or Whorled Simple Leaves

1 Plants parasitic, growing on limbs of deciduous trees; stems stout, jointed, dull green, brittle at base; leaves thick, fleshy, entire........American mistletoe,
Phoradendron flavescens (p. 166)
1' Plants rooted in soil:
  2 Stems climbing, creeping, or low (less than 15 cm high):
    3 Stems climbing or sprawling; vines:
      4 Aerial rootlets present; stems high climbing; leaf margins coarsely toothed to entire; petioles more than 2 cm long.... southern decumaria,
Decumaria barbara (p. 185)
      4' Aerial rootlets absent; stems twining or trailing, often forming dense tangles; leaf margins usually entire but occasionally lobed; petioles less than 2 cm long....................honeysuckle, *Lonicera* (p. 353)
    3' Stems low growing, creeping, or prostrate:
      5 Leaves entire:
        6 Leaf blades more than 2 cm long; sap milky in young stems; bark not shreddy .... periwinkle, *Vinca minor* (p. 341)
        6' Leaf blades mostly less than 15 mm long; sap not milky; bark of older stems papery and shreddy .... Allegany sand-myrtle,
Leiophyllum buxifolium var. *prostratum* (p. 313)
      5' Leaves sharply or finely toothed:
        7 Stems red or reddish brown, roundish; leaves apparently crowded into whorls, the blades variegated on upper side with a white or grayish band along midrib and larger veins
.... pipsissewa, *Chimaphila maculata* (p. 304)
        7' Stems green, four-sided; leaves strictly opposite, not variegated
.... running strawberry-bush, *Euonymus obovatus* (p. 259)
  2' Stems not low and not climbing, creeping, or prostrate:
    8 Leaves palmately veined, with more than one large vein arising from or near base of blade:
      9 Leaf blades very large and broad, usually more than 15 cm long and often more than 30 cm long, the margins entire or occasionally lobed:
        10 Leaves two at node (opposite), soft-velvety to touch; pith chambered or twigs hollow .... royal paulownia,
Paulownia tomentosa (p. 343)
        10' Leaves mostly three at node (whorled), not soft-velvety; pith solid
.... catalpa, *Catalpa* (p. 346)

9′ Leaf blades smaller, not more than 15 cm long, or if more than 15 cm, the margin toothed and lobed:

  11 Leaves only occasionally opposite and usually alternate, some lobed but many not lobed, rough on upper surface, softly hairy beneath; sap milky (best seen at end of broken petiole) .... paper-mulberry, *Broussonetia papyrifera* (p. 161)

  11′ Leaves regularly opposite, not rough on upper surface; sap not milky:

    12 Leaves lobed; terminal bud present:

      13 Petioles mostly less than 3 cm long; leaves three-lobed or sometimes not lobed, with minute black or reddish dots on lower side (visible under magnifying lens); small stipules often present on petiole .... mapleleaf viburnum, *Viburnum acerifolium* (p. 360)

      13′ Petioles mostly more than 3 cm long; leaves three- to five-lobed, without black or reddish dots on lower side and without stipules (leafy appendages commonly present in *A. nigrum*)......................... maple, *Acer* (p. 263)

    12′ Leaves entire or finely toothed; terminal bud absent .... mock-orange, *Philadelphus* (p. 186)

8′ Leaves pinnately veined, with only one primary vein or midrib:

  14 Leaves entire, or mostly entire but a few leaves irregularly toothed or wavy margined:

    15 Leaves dotted on upper surface with minute but distinct translucent glands; stems more or less two-angled below nodes .... St. Johnswort, *Hypericum* (p. 292)

    15′ Leaves without translucent glands:

      16 Leaves, mostly or all in whorls of three or four at node:

        17 Leaves leathery, 3–7 cm long; tending to cluster near end of current season's growth.................... lambkill, *Kalmia angustifolia* (p. 315)

        17′ Leaves not leathery, 6–16 cm long, scattered in pairs or whorls along current season's growth .... buttonbush, *Cephalanthus occidentalis* (p. 350)

      16′ Leaves two at node (opposite):

        18 Leaves mostly less than 4 cm long:

          19 Petiole bases of opposite leaves meeting around twig or connected by a definite ridge or line:

            20 Pith hollow between nodes; bundle scars three ....honeysuckle, *Lonicera* (p. 353)

            20′ Pith solid; bundle scar one .... coralberry, *Symphoricarpos orbiculatus* (p. 357)

19′ Petiole bases of opposite leaves not meeting and not
   connected .... privet, *Ligustrum* (p. 339)
18′ Leaves mostly more than 4 cm long:
   21 Leaves with lateral veins strongly curving forward, nearly
      parallel to leaf margin, upper veins meeting in apex
      ....dogwood, *Cornus* (p. 299)
   21′ Leaves with lateral veins not as above:
      22 Lateral buds concealed:
         23 Foliage aromatic when crushed; lateral buds
            nearly or completely covered by base of leaf
            petioles; leaves regularly opposite
            .... sweetshrub, *Calycanthus* (p. 179)
         23′ Foliage not aromatic when crushed; lateral buds
            buried in bark or appearing as pimplelike
            swellings about juncture of leaf petiole and twig;
            leaves opposite or in threes or fours
            ..... buttonbush, *Cephalanthus occidentalis* (p. 350)
      22′ Lateral buds clearly evident:
         24 Leaves heart shaped, especially at base, and
            tapering to a slender point .... common lilac,
                                        Syringa vulgaris (p. 339)
         24′ Leaves not heart shaped:
            25 Bud scales persisting where current year's
               growth meets growth of the previous year
               .... honeysuckle, *Lonicera* (p. 353)
            25′ Bud scales not persisting as above:
               26 Lower surface of leaves rusty dotted or
                  scurfy; petiole bases at each node meeting
                  or connected by transverse line
                  .... viburnum, *Viburnum* (p. 358)
               26′ Lower surface of leaves glabrous or hairy,
                  not rusty dotted or scurfy; petiole bases
                  not meeting or connected:
                  27 Leaves large, 8–20 cm long; twigs
                     moderate to stout, more than 2 mm
                     across .... fringetree, *Chionanthus*
                                        virginicus (p. 338)
                  27′ Leaves smaller, 3–8 cm long; twigs
                     slender, less than 2 mm across
                     .... privet, *Ligustrum* (p. 339)
14′ Leaves distinctly toothed:
   28 Sap milky; leaves usually alternate, often lobed .... paper-mulberry,
                                        Broussonetia papyrifera (p. 161)

28' Sap not milky; leaves regularly opposite:

    29 Petiole bases of opposing leaves meeting around twig or joined by a transverse line:

        30 Terminal bud naked, long stalked, scurfy, consisting of rudimentary leaves serving as scales . . . . . . .hobblebush, *Viburnum alnifolium* (p. 361)

        30' Terminal bud scaly, with one or more pairs of scales visible:

            31 Bud scales from last year's buds persisting at base of young twigs:

                32 Leaves mostly ovate; petioles 2–11 cm long; older bark loose, papery . . . . wild hydrangea, *Hydrangea arborescens* (p. 188)

                32' Leaves mostly lanceolate; petioles less than 1 cm long or absent; older bark not loose or papery . . . . bush-honeysuckle, *Diervilla* (p. 352)

            31' Bud scales from last year's buds not persisting at base of young twigs . . . . .viburnum, *Viburnum* (p. 358)

    29' Petiole bases of opposing leaves not meeting and not joined by transverse line:

        33 Twigs green, four-angled, squarish in cross section; leaves glabrous beneath or hairy only along veins . . . . .euonymus, *Euonymus* (p. 258)

        33' Twigs grayish brown, circular or nearly so in cross section; leaves white-woolly beneath . . . . beautyberry, *Callicarpa americana* (p. 341)

Key III. Plants with Opposite or Whorled Compound Leaves

1 Leaves palmately compound; leaflets five to seven, all radiating from top of petiole . . . . buckeye, *Aesculus* (p. 272)
1' Leaves pinnately compound, arranged in two rows along petiole:
  2 Stems climbing, sprawling, or trailing; vines:
    3 Stems with six longitudinal ridges and therefore six-angled in cross section; plants climbing by means of twining petioles
    . . . . clematis, *Clematis* (p. 168)
    3' Stems not ridged and angled as above:
     4 Stems with aerial rootlets at the nodes; leaflets seven to fifteen, toothed . . . . trumpet creeper, *Campsis radicans* (p. 345)
     4' Stems without aerial rootlets; leaflets two, entire, with a branched tendril between them, clinging by small disks. . . . . . . cross vine, *Bignonia capreolata* (p. 345)
  2' Stems self-supporting; trees and shrubs:
    5 Petiole bases of opposing leaves meeting around twig or connected by a transverse line or ridge:
     6 Leaflets three to five, with a few large, irregular teeth; twigs usually bright green and glaucous. . . . . boxelder, *Acer negundo* (p. 265)
     6' Leaflets five to eleven, sharply and regularly toothed; twigs greenish brown or reddish brown to gray, not glaucous. . . . . elder, *Sambucus* (p. 366)
    5' Petiole bases of opposing leaves not meeting or connected by a transverse line or ridge (but stipules or stipule scars present in *Staphylea*):
     7 Leaflets three; older twigs white striped; terminal bud lacking, twigs commonly ending with topmost pair of greenish lateral buds
     . . . . bladdernut, *Staphylea trifolia* (p. 263)
     7' Leaflets five or more; older twigs not striped; terminal bud present, large, blunt, rusty brown to almost black . . . . ash, *Fraxinus* (p. 335)

Key IV. Plants with Alternate Simple Leaves

(Some leaves may be crowded toward tips of twigs or on alternate spurs)

Index to Main Divisions of the Key

Read downward until you reach the first appropriate heading, then follow the key as shown under number and page. **These main divisions of the key are in boldface type.**

1 **Plants with stems climbing or scrambling; vines:**
  2 Stems with tendrils:
    3 Tendrils borne near base of leaf petioles; prickles usually present; leaves entire or nearly so, leathery......... greenbrier, catbrier, *Smilax* (p. 104)
    3′ Tendrils borne on stem opposite leaves; prickles absent; leaves toothed and often lobed, not leathery:
      4 Bark without lenticels, shredding into long strips on older stems; pith with woody partition at each node...................... grape, *Vitis* (except *V. rotundifolia*) (p. 278)
      4′ Bark dotted with warty lenticels, not shredding; pith without woody partitions at the nodes;
        5 Pith white; leaves triangular to heart shaped ... heartleaf ampelopsis, *Ampelopsis cordata* (p. 283)

      5' Pith brown in older stems; leaves almost circular in general outline
          .... muscadine, *Vitis rotundifolia* (p. 279)
2' Stems without tendrils:
    6 Plants climbing by means of aerial rootlets; leaves thick and leathery,
       mostly lobed........................ English ivy, *Hedera helix* (p. 295)
    6' Plants twining or scrambling; aerial rootlets absent:
      7 Leaves toothed, the teeth rounded or incurved
         .... bittersweet, *Celastrus* (p. 260)
      7' Leaves not toothed:
        8 Leaves pinnately veined, variable in shape, entire or deeply lobed
           near base of blade; foliage with strong disagreeable odor when
           rubbed between fingers; pith spongy or stems hollow
           .... bitter nightshade, *Solanum dulcamara* (p. 342)
        8' Leaves palmately veined; foliage without strong or disagreeable
           odor; pith solid:
          9 Petioles attached to underside of blades a short distance inside
             margin .... moonseed, *Menispermum canadense* (p. 172)
          9' Petioles attached to blades at margin:
            10 Leaves variable, ranging from ovate or broadly triangular to
                three-lobed, all on same stem, softly hairy beneath; blade
                margin finely ciliate .... coralbeads, *Cocculus*
                                           carolinus (p. 172)
            10' Leaves nearly circular but deeply heart shaped at base, never
                lobed, essentially glabrous; blade margin not ciliate
                .... Dutchman's-pipe, *Aristolochia durior* (p. 166)
1' Plants with stems not climbing or scrambling; not vines:
  11 **Stems low, creeping, prostrate, or mat forming, often scarcely woody:**
      12 Leaves small, 5–18 mm long, entire............. cranberry, *Vaccinium*
                                     macrocarpon (p. 326)
      12' Leaves larger, more than 2 cm long:
        13 Leaves palmately veined and usually lobed; stems frequently
           climbing, with aerial rootlets...... English ivy, *Hedera helix* (p. 295)
        13' Leaves pinnately veined, not lobed; stems not climbing, without
           aerial rootlets:
          14 Leaves entire; stems covered with coarse reddish or brownish
             hairs................. trailing-arbutus, *Epigaea repens* (p. 320)
          14' Leaves toothed; stems hairy or without hairs, but not covered
             with coarse reddish or brownish hairs;
            15 Leaves lanceolate, striped with white along principal
                veins, with no odor of wintergreen
                .... pipsissewa, *Chimaphila maculata* (p. 304)
            15' Leaves broader than lanceolate, not striped with white,
                aromatic with odor of wintergreen.... aromatic wintergreen,
                           Gaultheria procumbens (p. 321)

11′ Stems mostly erect or ascending, rarely low, sprawling, or reclining; trees and shrubs:
  16 **Leaves palmately veined:**
    17 Plants armed with thorns or prickles:
      18 Twigs with prominent ridges extending downward from nodes; bark loosening and shredding on older twigs; armed with prickles at some or all nodes
        .... gooseberry, *Ribes* (p. 189)
      18′ Twigs without prominent ridges; outer bark remaining close on older twigs; armed with thorns or thorn-tipped short growths:
        19 Thorns regularly present on young twigs as well as on older twigs, not or rarely bearing leaves or buds; bud present at base of thorn ...... hawthorn, *Crataegus* (p. 204)
        19′ Thorns not present on young twigs but only on older twigs, commonly bearing leaves or buds; bud lacking at base of thorn.... sweet crab apple, *Malus coronaria* (p. 208)
    17′ Plants not armed:
      20 Sap milky (as at base of a broken petiole); leaves toothed and often lobed:
        21 Petioles 2–5 cm long; leaves glabrous or hairy beneath but not velvety, regularly alternate
        .... mulberry, *Morus* (p. 162)
        21′ Petioles 5–10 cm long; leaves velvety-hairy beneath, sometimes opposite .... paper-mulberry, *Broussonetia papyrifera* (p. 161)
20′ Sap not milky:
  22 Base of petioles hollow, covering the buds; leaves three to five lobed, margined with large teeth .... sycamore, *Platanus occidentalis* (p. 196)
  22′ Base of petioles not hollow:
    23 Leaves with two or more large veins branching short distance above point where petiole joins blade; leaves entire or two- or three-lobed
    .... sassafras, *Sassafras albidum* (p. 182)
    23′ Leaves with three to five nearly equal large veins arising at point where petiole joins blade:
      24 Leaves densely coated with woolly hairs beneath:
        25 Twigs glabrous; leaves sharply and regularly toothed
        .... white basswood, *Tilia heterophylla* (p. 286)
        25′ Twigs whitish-woolly, at least near tip; leaves coarsely and irregularly toothed, occasionally lobed .... white poplar, *Populus alba* (p. 113)
      24′ Leaves not densely coated with woolly hairs beneath:
        26 Leaves, or most of them, lobed:

27 Twigs bristly, soft wooded; petioles densely glandular-hairy; bark loose and shreddy.........flowering raspberry, *Rubus odoratus* (p. 213)

27' Twigs not bristly, definitely woody:

   28 Leaves deeply five- to seven-lobed and star shaped, the lobes finely and regularly toothed .... sweetgum, *Liquidambar styraciflua* (p. 192)

   28' Leaves not deeply five- to seven-lobed and not star shaped:

      29 Stipules evident, at least near tip of twig, or their scars present; leaves appearing regularly alternate:

         30 Older bark tight, not peeling; leaves with large veins prominently raised on both sides of blade............. rose-of-Sharon, *Hibiscus syriacus* (p. 289)

         30' Older bark loose, peeling in thin strips; leaves with large veins not appearing raised on upper side of blade, but raised only on lower side............. ninebark, *Physocarpus opulifolius* (p. 198)

      29' Stipules and stipule scars lacking; leaves alternate on long growths, seemingly clustered on spur twigs .... gooseberry, currant, *Ribes* (p. 189)

26' Leaves not lobed:

   31 Petioles less than 2 cm long:

      32 Leaves with the left and right sides equal or nearly so at base; pith continuous; low shrub..... New Jersey tea, *Ceanothus americanus* (p. 276)

      32' Leaves unequal at base, one side distinctly larger than the other; pith finely chambered; trees ....hackberry, *Celtis* (p. 158)

   31' Petioles more than 2 cm long:

      33 Leaves entire, heart shaped to nearly circular .... redbud, *Cercis canadensis* (p. 230)

      33' Leaves toothed:

         34 Leaves usually unsymmetrical at base, the left and right sides unequal in size; pith circular in cross section ....basswood, *Tilia* (p. 286)

         34' Leaves symmetrical at base, the left and right sides about equal in size; pith angled in cross section ..... cottonwood, poplar, *Populus* (p. 112)

16′ Leaves pinnately veined:

  35 **Twigs with thorns or prickles or with spiny leaves:**

    36 Leaves thick and leathery, with a few large spiny teeth
      .... American holly, *Ilex opaca* (p. 255)

    36′ Leaves not thick and leathery, without spiny teeth:

      37 Plants with thorns present on young twigs as well as on older
        growth, the thorns not or rarely bearing leaves:

        38 Twigs with thorns, simple or three-pronged, just below
          clusters of leaves; inner bark and wood yellow
          .... barberry, *Berberis* (p. 170)

        38′ Twigs with thorns in axils of leaves:

          39 Leaves entire; sap milky .... Osage-orange, *Maclura
            pomifera* (p. 161)

          39′ Leaves toothed and often lobed; sap not milky
            .... hawthorn, *Crataegus* (p. 204)

      37′ Plants with short growths (spurs), the latter often thornlike or
        thorn tipped, rarely present on young twigs but only on older
        growth, commonly bearing leaves:

        40 Lower surface of leaves uniformly silvery; twigs and buds
          covered with silvery and sometimes brownish scaly particles
          .... autumn-olive, *Elaeagnus umbellata* (p. 294)

        40′ Lower surface of leaves not silvery; twigs and buds not as
          above:

          41 Leaves, or some of them, lobed. .... crab apple,
            Malus (p. 208)

          41′ Leaves not lobed, but variously toothed or entire:

            42 Leaves with (*a*) sharp, forward-pointing teeth and a
              wrinkled-veiny upper surface, or (*b*) red petioles
              bearing one or two pinhead-sized glands near base
              of blade; terminal bud lacking .... plum, *Prunus*
              (p. 223)

            42′ Leaves not as above in all respects; petioles not red;
              terminal bud present:

              43 Leaves all finely and evenly blunt toothed,
                lustrous or waxy on upper surface; terminal bud
                conical, pointed, brownish, glabrous
                .... pear, *Pyrus communis* (p. 211)

              43′ Leaves unevenly sharp or blunt toothed, or some
                leaves entire, rarely lustrous or waxy on upper
                surface; terminal bud usually ovoid, blunt, more
                or less hairy, or if pointed and glabrous, then the
                scales reddish .... crab apple, *Malus* (p. 208)

35′ Twigs without thorns, prickles, or spiny leaves:

  44 **Twigs nearly or completely encircled at the nodes by fine lines (stipule scars) or, during period of growth, by leaflike sheathing stipules:**

    45 Leaves regularly and sharply toothed, with lateral veins straight and parallel, a principal vein reaching each tooth; stipule scars not quite meeting around twig . . . . . . American beech, *Fagus grandifolia* (p. 140)

    45′ Leaves not toothed as above; lateral veins not straight and parallel; stipule scars forming a complete ring around twig:

      46 Leaves with margins entire or two-lobed at base of blade . . . . magnolia, *Magnolia* (p. 174)

      46′ Leaves with three to six lobes:

        47 Leaves three- or five-lobed and with pointed apex; buds covered by the hollow base of leaf petioles, their presence not evident unless leaf is removed . . . . sycamore, *Platanus occidentalis* (p. 196)

        47′ Leaves four- or six-lobed and with broad, shallowly notched apex; buds flattened, blunt, with two valvate scales . . . . yellow-poplar, *Liriodendron tulipifera* (p. 177)

  44′ Twigs not encircled at nodes by stipule scars or stipules:

    48 **Leaf buds almost or entirely concealed or apparently lacking:**

      49 Base of leaf petioles hollow, forming a hood over the buds; twigs swollen at nodes, with tough bark, difficult to break by hand . . . . leatherwood, *Dirca palustris* (p. 294)

      49′ Base of leaf petioles not hollow; twigs not as above:

        50 Leaves thick, leathery, entire; leaf buds concealed between twig and petiole . . . . mountain-laurel, *Kalmia latifolia* (p. 314)

        50′ Leaves thin, not leathery, toothed; buds concealed by upturned wings along petioles . . . . mountain camelia, *Stewartia ovata* (p. 290)

    48′ Leaf buds visible on well-developed growth:

      51 **Leaves all or mostly lobed:**

        52 Stems nearly herbaceous, woody only at base, usually climbing or trailing; leaves mostly with pair of basal lobes but variable, some without lobes and a few compound . . . . bitter nightshade, *Solanum dulcamara* (p. 342)

        52′ Stems definitely woody; erect trees or shrubs:

          53 Leaves mostly four-lobed, very broadly and shallowly notched at apex; buds large, shaped like a duck's bill, with valvate scales . . . . yellow-poplar, *Liriodendron tulipifera* (p. 177)

          53′ Leaves not four-lobed; bud scales not valvate:

    54 Buds three or more in a cluster at tip of twigs, each
       bud with numerous scales.......oak, *Quercus* (p. 142)
    54′ Buds fewer than three, sometimes none, at tip of
       twigs:
      55 Foliage spicy-aromatic or fragrant when
        crushed; sharply pointed spur twigs absent:
        56 Leaves regularly cut or divided into fifteen to
          twenty-five lobes, dotted with resin droplets
          on both surfaces; low shrub, often in
          colonies.............. sweet-fern, *Comptonia*
          peregrina (p. 117)
        56′ Leaves two- or three-lobed or entire, all
          three types commonly present on the same
          plant, not resin dotted; large shrub or tree
          .... sassafras, *Sassafras albidum* (p. 182)
      55′ Foliage not spicy-aromatic or fragrant; sharply
        pointed spur twigs often present........crab apple,
        Malus (p. 208)

51′ Leaves not lobed:
  **57 Leaves entire or mostly entire, but some leaves minutely or irregularly
  toothed or wavy margined:**
    58 Leaves mostly more than 15 cm long:
      59 Leaves thick and leathery, those of the previous year persisting;
        terminal bud scaly with many overlapping scales
        ...... rosebay rhododendron, *Rhododendron maximum* (p. 306)
      59′ Leaves not thick and leathery, those of the previous year not
        persisting:
        60 Terminal bud naked, hairy, dark reddish brown; twigs not
          aromatic................. pawpaw, *Asimina triloba* (p. 180)
        60′ Terminal bud covered with single scale, or if densely hairy
          and appearing naked, not dark reddish brown; twigs
          aromatic......................magnolia, *Magnolia* (p. 174)
    58′ Leaves mostly less than 15 cm long:
      61 Leaves rather leathery, those of the previous year persisting on
        older twigs:
        62 Leaves small, less than 2 cm long; plants low, less than 30
          cm high when mature, densely branched.........sand-myrtle,
          Leiophyllum buxifolium (p. 313)
        62′ Leaves larger, more than 4 cm long; plants medium to tall
          shrubs .... rhododendron, *Rhododendron* (p. 306)
      61′ Leaves not leathery, or if so those of the previous year not
        persisting:
      63 Lower surface of leaves evenly dotted with silvery scales or

scurf; young twigs yellowish brown or silvery; leaf margin
often undulating but not definitely toothed
.... autumn-olive, *Elaeagnus umbellata* (p. 294)
63′ Lower surface of leaves not dotted with silvery scales or
scurf:
    64 Leaves and twigs spicy-aromatic:
        65 Leaves with three primary veins from near base,
often two- or three-lobed, very unevenly spaced
along twig; twigs commonly branching in first
season, the terminal twig surpassed in length by the
branch just below it .... sassafras, *Sassafras
albidum* (p. 182)
        65′ Leaves with single primary vein (midrib), never
lobed, evenly spaced along twig; twigs rarely
branching in first season .... spicebush, *Lindera
benzoin* (p. 182)
64′ Leaves and twigs not spicy-aromatic:
  66 Twigs and leaf petioles with milky sap; sharp thorns often present, a
small bud at the side of each thorn .... Osage-orange, *Maclura
pomifera* (p. 161)
66′ Twigs and leaf petioles without milky sap:
  67 Base of petioles enlarged and hollow, covering buds like a hood;
twigs appearing jointed, with brittle wood but very tough bark
.... leatherwood, *Dirca palustris* (p. 294)
67′ Base of petioles not enlarged and hollow:
  68 Leaves with veins curving strongly forward, nearly parallel to
margin, the topmost veins meeting at apex; twigs often
branching in first year, the branches surpassing the terminal
twig in length .... alternate-leaf dogwood, *Cornus
alternifolia* (p. 299)
68′ Leaves with veins not as above:
  69 Tip of twigs bearing three or more buds in a cluster;
branches often grouped, appearing whorled:
    70 Pith chambered, the chambers empty; foliage with
sweetish taste ..... sweetleaf, *Symplocos tinctoria* (p. 332)
    70′ Pith continuous; foliage not sweetish:
      71 Pith four- or five-angled in cross section; leaves
ending in a hairlike bristle; bundle scars many
.... shingle oak, *Quercus imbricaria* (p. 146)
      71′ Pith rounded in cross section; leaves not ending in a
hairlike bristle; bundle scar one:
        72 Lower surface of leaves with small, fringed
glands along midrib, usually visible without

magnifying lens .... minnie-bush, *Menziesia pilosa* (p. 312)

72′ Lower surface of leaves without glands along midrib............rhododendron, *Rhododendron* (p. 304)

69′ Tip of twigs not bearing three or more buds in a cluster:

73 Leaves dotted with yellow resin droplets on lower surface; twigs slender, often pink or green............ huckleberry, *Gaylussacia* (p. 321)

73′ Leaves without resin droplets as above:

74 Buds naked, clothed with reddish brown hairs; leaves dark green and lustrous on upper side .... glossy buckthorn, *Rhamnus frangula* (p. 276)

74′ Buds scaly with one or more scales:

75 Buds covered with a single caplike scale; leaf margin entire, wavy, or with scattered blunt teeth .... willow, *Salix* (p. 108)

75′ Buds covered with two or more scales:

76 Leaves mostly not more than 7 cm long:

77 Twigs bright green or yellowish green and speckled with warty dots (sometimes obscured by hairs), or else buds nearly globose and widely divergent .... blueberry, *Vaccinium* (p. 324)

77′ Twigs typically yellow, not speckled with warty dots; buds oblong, flattened, sharp pointed, usually pressed close to twig ....male-berry, *Lyonia ligustrina* (p. 316)

76′ Leaves mostly more than 7 cm long:

78 Leaves very soft to touch, prominently veiny; twigs commonly self-pruned, leaving twig scars; buds large, greenish, often borne in irregular groups or clusters ....buffalo nut, *Pyrularia pubera* (p. 165)

78′ Leaves not so soft or veiny; twigs not self-pruned; buds not borne in irregular groups or clusters:

79 Leaves mostly wedge shaped at base; terminal bud present:

80 Leaves sweet to taste, often faintly toothed; bundle scar one ....sweetleaf, *Symplocos tinctoria* (p. 332)

80′ Leaves not sweet to taste, rarely with a few coarse teeth; bundle scars three ...... black tupelo, *Nyssa sylvatica* (p. 298)

79′ Leaves mostly rounded at base; terminal bud absent; bundle scar one .... persimmon, *Diospyros virginiana* (p. 331)

57' Leaves not entire:
  81 **Leaves coarsely toothed, with only one or two teeth per centimeter of margin:**
     82 Buds clearly stalked when mature:
        83 Leaves with the lowest pair of lateral veins forming margin of blade for 2–10 mm on each side of midrib; leaf base symmetrical or nearly so . . . . large fothergilla, *Fothergilla major* (p. 193)
        83' Leaves with the lowest pair of lateral veins not forming margin of blade as above; leaf base unsymmetrical, left and right sides decidedly unequal in size . . . . witch-hazel, *Hamamelis virginiana* (p. 192)
     82' Buds not stalked:
        84 Leaves broadly ovate, nearly circular, or triangular; bark of twigs very bitter . . . . cottonwood, poplar, *Populus* (p. 112)
        84' Leaves not as above, but much longer than wide; bark of twigs not markedly bitter:
           85 Buds three or more in cluster at tip of twig . . . . . . oak, *Quercus* (p. 142)
           85' Buds one or none at tip of twig:
              86 Leaves with each tooth ending in a hair or bristle; buds ovoid, with two or three visible scales
                 . . . . chestnut, chinkapin, *Castanea* (p. 140)
              86' Leaves with teeth not ending in a hair or bristle; buds lance shaped, with ten or more visible scales
                 . . . . American beech, *Fagus grandifolia* (p. 140)
  81' **Leaves not so coarsely toothed, usually with three to many teeth per centimeter of margin** (this section includes plants having leaves finely toothed but the teeth widely spaced):
     87 Lateral veins clearly extending nearly straight to leaf margin and ending in the teeth:
        88 Twigs with many short spurs regularly bearing two leaves and ending in a terminal bud . . . . birch, *Betula* (p. 129)
        88' Twigs without short spurs as above, or if spur twigs present, not regularly bearing two leaves:
           89 Buds distinctly stalked; fruiting catkins woody, conelike, long persistent . . . . alder, *Alnus* (p. 133)
           89' Buds not distinctly stalked:
              90 Leaves decidedly unsymmetrical, the left and right sides unequal, particularly at base . . . . elm, *Ulmus* (p. 155)
              90' Leaves symmetrical, the two sides equal or nearly so at base:

91   Leaves usually tapering at base; bundle scar one
. . . . spiraea, *Spiraea* (p. 199)

91'   Leaves usually rounded or heart shaped at base;
bundle scars three or more:

   92   Leaves with fewer than ten prominent veins on
each side of midrib:

     93   Buds reddish purple, gummy, often short
stalked; fruiting structures woody, conelike,
in clusters, those of previous year persisting
all summer on mature plants
. . . . mountain alder, *Alnus crispa* (p. 134)

     93'   Buds yellowish brown to reddish brown, not
gummy or stalked; fruit a nut in close-fitting
husk, mature in late summer or early
autumn . . . . hazelnut, *Corylus* (p. 137)

   92'   Leaves with ten or more prominent veins on
each side of midrib:

     94   Leaves single toothed, short pointed or blunt
at apex; terminal bud present
. . . . roundleaf serviceberry,
Amelanchier sanguinea (p. 206)

     94'   Leaves sharply double toothed, long pointed
at apex; terminal bud absent:

       95   Leaves with lateral veins clearly
branched near margin and ending in all
or nearly all the teeth; trunk and larger
branches with rough, grayish brown
bark, the outer layer often in loose
vertical strips. . . . . . . eastern hophornbeam,
Ostrya virginiana (p. 137)

       95'   Leaves with lateral veins unbranched
and ending only in the larger teeth or, if
branched, the smaller veins indistinct
and scarcely traceable unless magnifying
lens is used; trunk and larger branches
fluted, appearing "muscular," with
smooth, dark gray bark
. . . . American hornbeam,
Carpinus caroliniana (p. 135)

87' Lateral veins curving before reaching leaf margin, or divided into a network
of small veinlets, or indistinct and not easily traced:

  96   Buds naked or so densely hairy that scales are obscured:

   97   Leaf petioles with two broad wings almost concealing lateral buds;

buds silvery-silky .... mountain camelia, *Stewartia ovata* (p. 290)

97' Leaf petioles with only narrow wings or without wings:

    98 Young twigs yellowish scurfy or rusty scurfy; bark of older twigs reddish brown, separating into loose strips or squarish pieces .... mountain pepperbush, *Clethra acuminata* (p. 303)

    98' Young twigs not yellowish scurfy or rusty scurfy; bark of older twigs not as above:

        99 Primary lateral veins of leaves curving evenly and strongly toward apex, finally ending in the margin; terminal bud present, naked...................... Carolina buckthorn, *Rhamnus caroliniana* (p. 275)

        99' Primary lateral veins not curving as above and not ending in the margin:

            100 Petioles more than 15 mm long; leaves usually less than twice as long as wide; terminal bud present, several-scaled .... apple, *Malus sylvestris* (p. 208)

            100' Petioles less than 15 mm long; leaves usually three times or more as long as wide; terminal bud absent; lateral buds single scaled .... willow, *Salix* (p. 108)

96' Buds scaly with one or more clearly visible scales:

    101 Buds covered by a single caplike scale; bark very bitter; twigs usually slender, often brittle at base, easily broken off ..... willow, *Salix* (p. 108)

    101' Buds covered by two or more visible scales:

        102 Leaves with minute dark glands along midrib on upper side .... chokeberry, *Aronia* (p. 202)

        102' Leaves without glands on upper midrib:

            103 Twigs with chambered pith:

                104 Bark of twigs soon becoming silky-shreddy; older bark striped with white; bundle scar one ....Carolina silverbell, *Halesia carolina* (p. 333)

                104' Bark of twigs tight, not silky-shreddy; older bark not white striped:

                    105 Twigs brown or grayish brown; leaves thick, rather leathery, sweet tasting; bundle scar one .... sweetleaf, *Symplocos tinctoria* (p. 332)

                    105' Twigs green or sometimes reddish green; leaves thin, not sweet tasting; bundle scars three .... sweet-spires, *Itea virginica* (p. 185)

            103' Twigs with solid or spongy pith:

                106 Pith spongy; buds silvery, 2–7 mm long, lateral buds commonly superposed, a smaller bud at base of a larger one; bundle scar one, large, at top of leaf scar .... stewartia, *Stewartia* (p. 289)

106′ Pith solid; buds and bundle scars not as above in all respects:

  107 Leaf scars with three bundle scars (or three vascular bundles evident if leaf petiole is cut cleanly across at base); terminal bud present:

    108 Leaves about as broad as long, usually somewhat triangular; lowest scale of lateral buds centered directly above leaf scar .... cottonwood, poplar, *Populus* (p. 112)

    108′ Leaves distinctly longer than broad; lowest scale of lateral buds to one side of leaf scar:

      109 Leaves mostly heart shaped at base; terminal bud long and slender, its length more than four times the width of bud at base, sharply pointed, with about six pinkish or greenish scales visible......... serviceberry, *Amelanchier* (p. 206)

      109′ Leaves rounded to wedge shaped at base; terminal bud not as above, usually ovoid or conical:

        110 Twigs with strong odor of bitter almonds or crushed cherry pits if cut or broken; leaves mostly with pinhead-sized glands at or near juncture of petiole and blade ...... cherry, peach, *Prunus* (p. 223)

        110′ Twigs without strong odor; leaves without glands as above:

          111 Leaves on vigorous growth glabrous beneath, finely and regularly toothed; terminal bud conical, pointed, glabrous, brownish or greenish........... pear, *Pyrus communis* (p. 211)

          111′ Leaves on vigorous growth either densely woolly beneath or else with several lobes or lobelike teeth; terminal bud ovoid, blunt, more or less hairy or, if

pointed and glabrous, the
scales reddish
.... apple, *Malus* (p. 208)

107' Leaf scars with only one bundle scar (or, if petiole is cut cleanly across at
base, only one vascular bundle evident, appearing as a dot or straight or
curved line); terminal bud absent:

  112 Stipules very small, triangular, dark colored, commonly persistent or
their scars visible (under magnifying lens); buds small, nearly
globose, often superposed........................ holly, *Ilex* (p. 255)

  112' Stipules not as above or absent; buds solitary:

    113 Leaves thick, leathery:

      114 Lower surface of leaves dotted with black or reddish
glands................ fetter-bush, *Pieris floribunda* (p. 315)

      114' Lower surface of leaves not dotted as above:

        115 Leaves 2–7 cm long, minutely toothed or entire,
rounded or short pointed at apex
............sparkleberry, *Vaccinium arboreum* (p. 326)

        115' Leaves 5–13 cm long, finely toothed, tapering to long
point at apex ....dog-hobble, *Leucothoe
editorum* (p. 317)

    113' Leaves thin, not leathery:

      116 Leaves coarsely or doubly toothed; lateral veins curving
upward, but the larger ones extending to the teeth
..................................spiraea, *Spiraea* (p. 199)

      116' Leaves very finely or singly toothed; lateral veins either
indistinct or not reaching the teeth:

        117 Twigs yellow, yellowish brown, or gray, often mottled
with black; buds pinkish, with two valvate scales
.... maleberry, *Lyonia ligustrina* (p. 316)

        117' Twigs green or yellowish green to red, often green on
lower side, red on side facing light:

          118 Leaves large, 8–20 cm long, finely toothed, the
teeth usually confined to upper half of margin;
bundle scar crescent-shaped or U-shaped
.... sourwood, *Oxydendrum arboreum* (p. 320)

          118' Leaves smaller, mostly 3–10 cm long, the teeth
not confined to upper half of margin; bundle
scar not as above:

            119 Leaves clearly toothed; twigs not warty
dotted; leaf buds with three or four scales
visible; flower buds in long, narrow, dark
red to greenish clusters, the clusters mostly

3 cm long or more, each bud in axil of a
scalelike bract . . . . sweetbells, *Leucothoe*
(p. 317)

119′ Leaves finely and obscurely toothed; twigs
speckled with warty dots or else leaf buds
with two valvate scales; flower buds, if
present, not in long clusters
. . . . blueberry, *Vaccinium* (p. 324)

Key V. Plants with Alternate Compound Leaves

1 Fruits white or whitish, glabrous, berrylike, 4–7 mm across, in spreading or
drooping clusters; *poisonous to touch* . . . . poison-ivy, poison-sumac,
Toxicodendron (p. 252)

1' Fruits not as above or absent:

  2 Leaves with three leaflets, the central one having a stalk 1–2 cm long or
more, much longer than the stalks of lateral leaflets; margin of leaflets
entire or with a few large teeth or lobes; usually a vine with aerial rootlets,
climbing or trailing, but sometimes shrubby; *poisonous to touch*
   . . . . poison-ivy, *Toxicodendron radicans* (p. 252)

  2' Leaves with leaflets not as above in all respects:

    3 Stems armed with thorns, prickles, or stiff bristles:

      4 Stipules present, 5 mm long or more, attached for more than half
their length to the lower part of the petioles . . . . rose, *Rosa* (p. 219)

      4' Stipules not as above or absent:

        5 Plants with stout, usually long, branched thorns
         . . . . honeylocust, *Gleditsia triacanthos* (p. 232)

        5' Plants with unbranched thorns or with prickles and/or stiff
bristles:

          6 Stems with stout or bristlelike thorns in pairs at nodes:

            7 Buds clearly evident, red-hairy; leaflets showing translucent
dots if held to light . . . . prickly-ash, *Zanthoxylum
americanum* (p. 244)

            7' Buds nearly or entirely concealed under base of leaf petioles;
leaflets without translucent dots . . . . locust, *Robinia* (p. 238)

          6' Stems with weak to strong prickles scattered between nodes or
borne mostly in a transverse row just below nodes:

            8 Leaves twice or thrice compound, with many leaflets; base of
petioles clasping stems; erect shrub or small tree
             . . . . .devils-walkingstick, *Aralia spinosa* (p. 296)

            8' Leaves once compound, with three to five leaflets; base of
petioles not clasping stems; erect, arching, or trailing plants
             . . . . blackberry, dewberry, raspberry, *Rubus* (p. 212)

3' Stems not armed:

  9 Stems climbing or trailing; vines:

    10 Leaflets three; stems usually with aerial rootlets; *poisonous to touch*
     . . . . poison-ivy, *Toxicodendron radicans* (p. 252)

    10' Leaflets more than three; stems without aerial rootlets:

      11 Leaves palmately compound, with usually five leaflets, their
margin coarsely toothed; stems climbing by means of tendrils

usually ending in adhesive disks . . . . . Virginia creeper, *Parthenocissus quinquefolia* (p. 284)

11′ Leaves pinnately compound, with nine to fifteen leaflets, their margin entire; stems twining . . . . wisteria, *Wisteria* (p. 238)

9′ Stems not climbing or trailing; trees and shrubs:

12 Margin of leaflets entire or with only a few coarse teeth near base:

13 Leaflets entire except for one to four gland-bearing teeth at base; crushed foliage with disagreeable odor . . . . ailanthus, *Ailanthus altissima* (p. 245)

13′ Leaflets entire or essentially so throughout their length:

14 Petioles winged; twigs velvety, dotted with orange red lenticels . . . . shining sumac, *Rhus copallina* (p. 250)

14′ Petioles not winged:

15 Leaflets three:

16 Stems prominently ridged or angled, green: leaflets 15 mm long or less, some leaves simple . . . . Scotch broom, *Cytisus scoparius* (p. 235)

16′ Stems rounded, brown or reddish brown; leaflets 5–15 cm long . . . . hoptree, *Ptelea trifoliata* (p. 244)

15′ Leaflets more than three:

17 Lateral buds clearly visible:

18 Leaflets less than 5 cm long, marked with glandular dots beneath; terminal bud lacking . . . . indigobush, *Amorpha* (p. 236)

18′ Leaflets 5–12 cm long, not glandular dotted; terminal bud present, much larger than laterals; *poisonous to touch* . . . . poison-sumac, *Toxicodendron vernix* (p. 254)

17′ Lateral buds (*a*) partly or wholly concealed by hollow base of petioles or (*b*) sunken under petioles or in bark above petioles:

19 Leaves once compound:

20 Leaflets 6–10 cm long, mostly alternate along petioles; stipules lacking . . . . yellowwood, *Cladrastis kentukea* (p. 234)

20′ Leaflets 2–6 cm long, opposite along petioles; stipules spiny or prickly, or their scars present . . . . locust, *Robinia* (p. 238)

19′ Leaves, some or all, twice compound:

21 Leaflets unsymmetrical, with the midrib very close and parallel to one margin; twigs greenish . . . . silktree, *Albizia julibrissin* (p. 230)

21′ Leaflets symmetrical, with the midrib centrally placed; twigs brownish:

22 Leaflets less than 15 mm wide and faintly
    wavy toothed; twigs usually armed with
    branched thorns
    ... honeylocust, *Gleditsia triacanthos* (p. 232)
22′ Leaflets mostly 2 cm wide or more; twigs
    never armed........... Kentucky coffeetree,
                  Gymnocladus dioicus (p. 232)
12′ Margin of leaflets distinctly toothed or lobed:
  23 Leaves with stipules attached for more than half their length to base of
     petioles; stems usually prickly...................... rose, *Rosa* (p. 219)
  23′ Leaves with stipules not as above or absent:
    24 Stipules present and persistent, or if stipules falling early and
      leaving scars, buds red or purplish red:
      25 Leaflets three to five; terminal bud absent; stems usually prickly
        .... blackberry, dewberry, raspberry, *Rubus* (p. 212)
      25′ Leaflets thirteen to seventeen; terminal bud large and
        conspicuous; stems not prickly.......... American mountain-ash,
                  Sorbus americana (p. 205)
    24′ Stipules and stipule scars lacking:
      26 Leaflets three; central leaflet stalkless or nearly so, but
        appearing stalked because of long-tapering base
        .... fragrant sumac, *Rhus aromatica* (p. 249)
      26′ Leaflets mostly five or more:
        27 Base of petioles extending more than halfway around twig
          and clasping it; leaflets deeply toothed or lobed; roots and
          inner bark yellow .... yellowroot, *Xanthorhiza*
                  simplicissima (p. 168)
        27′ Base of petioles not as above; roots and inner bark not
          yellow:
          28 Sap milky, as at base of a broken petiole; terminal bud
            lacking.......................... sumac, *Rhus* (p. 248)
          28′ Sap not milky; terminal bud present, large and
            prominent:
            29 Pith chambered; leaflets eleven to twenty-three
              .... walnut, *Juglans* (p. 117)
            29′ Pith solid:
              30 Leaflets thirteen to seventeen; terminal bud dark
                red, gummy; leaf scars narrow, crescent-shaped
                or U-shaped............. American mountain-ash,
                    Sorbus americana (p. 205)
             30′ Leaflets five to eleven; terminal bud neither dark
               red nor gummy; leaf scars broad, shield shaped
               or three-lobed............hickory, *Carya* (p. 120)

Winter Keys

Key VI. Evergreen and Half-Evergreen Plants with Opposite or Whorled Leaves

1 Leaves compound; leaflets two, with a terminal tendril between them; vine
.................................cross vine, *Bignonia capreolata* (p. 345)
1' Leaves simple:
  2 Stems climbing or sprawling:
    3 Aerial rootlets present; pith spongy, greenish .... southern decumaria,
                                             Decumaria barbara (p. 185)
    3' Aerial rootlets absent; stems twining; pith hollow
.................................honeysuckle, *Lonicera* (p. 353)
  2' Stems not climbing or sprawling:
    4 Stems low, creeping or trailing, or forming mat or cushion:
      5 Leaves toothed, variegated with white along midrib and larger veins
.................................pipsissewa, *Chimaphila maculata* (p. 304)
      5' Leaves entire:
        6 Leaves less than 8 mm wide; plants definitely woody, dense,
          forming a cushion .... Allegany sand-myrtle,
                 Leiophyllum buxifolium var. *prostratum* (p. 313)
        6' Leaves more than 8 mm wide; plants scarcely woody, trailing, the
          flowering stems more or less upright:
          7 Leaves broadly ovate to nearly circular, 8–20 mm long, heart
            shaped to rounded at base, whitish along midrib
.................................partridgeberry, *Mitchella repens* (p. 350)
          7' Leaves elliptical to ovate, more than 20 mm long, wedge shaped
            at base, not whitish along midrib..............periwinkle, *Vinca*
                                  minor (p. 341)
  4' Stems erect:
    8 Leaves leathery, often in whorls of three; terminal bud lacking
.................................lambkill, *Kalmia angustifolia* (p. 315)
    8' Leaves not leathery, regularly opposite; terminal bud present, distinct
.................................privet, *Ligustrum* (p. 339)

Key VII. Evergreen and Half-Evergreen Plants with Alternate Leaves

1 Leaves compound, mostly with three leaflets; stems trailing, prickly and more or less bristly . . . . swamp dewberry, *Rubus hispidus* (p. 215)
1' Leaves simple:
   2 Plants high-climbing or thicket-forming vines, with tendrils borne on leaf petioles; stems armed with prickles, especially near base of plant
   . . . . greenbrier, catbrier, *Smilax* (p. 104)
   2' Plants not vines, not tendril bearing, and not armed:
      3 Leaves with hard, spiny teeth . . . . American holly, *Ilex opaca* (p. 255)
      3' Leaves without spiny teeth:
         4 Stems low, creeping or mat forming, or sending up short, erect stems from long underground stems:
            5 Leaves unevenly spaced, noticeably crowded toward tip of year's growth:
               6 Leaves lanceolate, with greenish white stripe along midrib; foliage without wintergreen odor and taste
               . . . . pipsissewa, *Chimaphila maculata* (p. 304)
               6' Leaves broader than lanceolate, uniformly green; foliage with wintergreen odor and taste . . . . . aromatic wintergreen, *Gaultheria procumbens* (p. 321)
            5' Leaves more or less evenly spaced along stems:
               7 Leaves small, 5–18 mm long; stems glabrous
               . . . . cranberry, *Vaccinium macrocarpon* (p. 326)
               7' Leaves larger, 3–8 cm long; stems bristly with reddish hairs
               . . . . trailing-arbutus, *Epigaea repens* (p. 320)
         4' Stems erect or ascending; trees and shrubs:
            8 Leaves small, mostly 5–15 mm long, entire; older bark shreddy
            . . . . sand-myrtle, *Leiophyllum buxifolium* (p. 313)
            8' Leaves larger, mostly 3 cm long or more:
               9 Leaves finely but clearly toothed, very long pointed; stems spreading and arching . . . . dog-hobble, *Leucothoe editorum* (p. 317)
               9' Leaves entire or only minutely toothed:
                  10 Leaves dotted beneath with numerous scales or depressions:
                     11 Buds clustered at tip of twigs; dotlike scales orange or reddish brown; leaves entire . . . . . Piedmont rhododendron, *Rhododendron minus* (p. 307)

      11′ Buds not clustered at tip of twigs; dotlike scales reddish
to black; leaves faintly toothed with bristle-tipped teeth
. . . . fetter-bush, *Pieris floribunda* (p. 315)

  10′ Leaves not dotted beneath with scales or depressions:

      12 Leaves crowded toward twig tips; leaf blades two or
more times as long as wide:

         13 Leaves 5–10 cm long; leaf buds two-scaled,
concealed between twig and petiole base on mature
growth. . . . . . mountain-laurel, *Kalmia latifolia* (p. 314)

        13′ Leaves averaging more than 10 cm long; leaf buds
with six or more scales visible, evident on mature
growth. . . . . rhododendron, *Rhododendron* (p. 304)

      12′ Leaves not crowded toward twig tips; leaf blades not
more than twice as long as wide
. . . . sparkleberry, *Vaccinium arboreum* (p. 326)

Key VIII. Deciduous Plants with Opposite or Whorled Leaf Scars

1 Plants climbing or trailing, not self-supporting:
  2 Stems climbing or twining; vines:
    3 Stems with aerial rootlets:
      4 Rootlets in two bands below the nodes: leaf scars shield shaped to half circular; buds glabrous, scaly.................... trumpet creeper, *Campsis radicans* (p. 345)
      4′ Rootlets not in two bands; leaf scars U-shaped; buds red-hairy .... southern decumaria, *Decumaria barbara* (p. 185)
    3′ Stems without aerial rootlets:
      5 Stems with six prominent ridges lengthwise, and thus appearing six-angled in cross section; plants climbing by means of tendril-like leaf petioles, which persist into winter..........clematis, *Clematis* (p. 168)
      5′ Stems without six prominent ridges as above:
        6 Leaf scars broadly shield shaped to triangular; bundle scar one, C-shaped, the open side up; plants climbing by coiling tendrils, the latter often persisting; stem, if carefully sliced across, showing four large pith rays forming a prominent cross .... cross vine, *Bignonia capreolata* (p. 345)
        6′ Leaf scars crescent shaped; bundle scars three; plants climbing by twining of stem around support; pith not as above ....honeysuckle, *Lonicera* (p. 353)
  2′ Stems trailing, rooting at nodes, usually four-angled, green ....running strawberry-bush, *Euonymus obovatus* (p. 260)
1′ Plants erect and self-supporting; trees or shrubs:
  7 Leaf scars with only one bundle scar, the scar appearing as a dot or a straight or curved line:
    8 Opposing leaf scars meeting around twig or joined by a transverse line or ridge:
      9 Twigs prominently two-edged and more or less flattened below nodes; leaf scars small, triangular; buds small but often developing into short, leafy stems in autumn or early winter .... St. Johnswort, *Hypericum* (p. 292)
      9′ Twigs not prominently two-edged and flattened below nodes; buds rarely developing into leafy stems as above:
        10 Buds partly or entirely concealed, sunken in bark above leaf scar; leaf scars nearly circular, readily visible, not raised on base of petioles, often whorled; bundle scar U-shaped; twigs usually

reddish and glossy; bark close . . . . buttonbush, *Cephalanthus occidentalis* (p. 350)

    10′ Buds small but fully exposed; leaf scars regularly opposite, very small, raised on persistent base of leaf petioles, often indistinct; bundle scar dotlike or obscure; older bark shreddy . . . . coralberry, *Symphoricarpos orbiculatus* (p. 357)

8′ Opposing leaf scars not meeting around twig and not joined by a transverse line or ridge:

    11 Buds naked, or the smaller ones covered with two nearly valvate scales; buds often stalked; bundle scar crescent shaped . . . . beautyberry, *Callicarpa americana* (p. 341)

    11′ Buds all scaly, not stalked:

        12 Twigs with two distinct pairs of vertical lines or ridges extending downward from nodes:

            13 Twigs bright green, or sometimes reddish on side exposed to sun, usually four-lined; terminal bud present, with three to five pairs of scales visible . . . . euonymus, *Euonymus* (p. 258)

            13′ Twigs olive green, brown, or gray, not four-lined; terminal bud usually lacking, twigs often ending with topmost pair of lateral buds, each with about four pairs of scales visible . . . . common lilac, *Syringa vulgaris* (p. 339)

        12′ Twigs without distinct vertical lines or ridges below nodes (faint lines may be present in *Chionanthus*):

            14 Twigs slender, less than 2 mm wide, often holding green leaves in winter; bundle scar extending across center of leaf scar . . . . privet, *Ligustrum* (p. 339)

            14′ Twigs moderate to stout, more than 2 mm wide; bundle scar consisting of many scars more or less fused:

                15 Terminal bud present; bundle scar U-shaped; bud scales light brown; lenticels prominent; leaf scars occasionally subopposite (one scar of a pair slightly higher than the other) or alternate . . . . fringetree, *Chionanthus virginicus* (p. 338)

                15′ Terminal bud absent; bundle scar a transverse line or broadly crescent shaped, directly under bud; bud scales green to reddish brown; leaf scars strictly opposite . . . . common lilac, *Syringa vulgaris* (p. 339)

7′ Leaf scars with three or more bundle scars:

  16 Lateral buds almost or entirely concealed or apparently lacking; that is, buds covered by leaf scars, or concealed behind persistent base of leaf petioles, or more or less buried in bark above leaf scars:

  17 Terminal bud present:

    18 Terminal bud naked, often made up of pleated, rudimentary

leaves, densely coated with scurfy, rusty to golden brown hairs; leaf scars broad, triangular to three-lobed
.... hobblebush, *Viburnum alnifolium* (p. 361)

18′ Terminal bud scaly and of two kinds: larger flower buds biscuit shaped, long stalked, and smaller leaf buds narrow, pointed, not stalked; leaf scars broadly U-shaped, raised on persistent base of leaf petioles, these later deciduous, exposing the true leaf scars
.... flowering dogwood, *Cornus florida* (p. 300)

17′ Terminal bud absent:

19 Twigs slender; lateral buds nearly or entirely buried under membrane covering the leaf scar and bursting through in spring; leaf scars half circular until membrane is broken by developing buds, then crescent shaped................ scentless mock-orange, *Philadelphus inodorus* (p. 186)

19′ Twigs stout; lateral buds buried or partially buried in bark above leaf scar; leaf scars large, circular to elliptical with depressed center:

20 Pith continuous; leaf scars usually three at a node (whorled)
.... catalpa, *Catalpa* (p. 346)

20′ Pith chambered or sometimes hollow; leaf scars two at a node
.... royal paulownia, *Paulownia tomentosa* (p. 343)

16′ Lateral buds visible:

21 Buds naked or at least without evident scales, densely hairy or woolly:

22 Buds large, long stalked, often appearing to consist of tightly folded rudimentary leaves, rusty brown, scurfy; leaf scars triangular or three-lobed, entirely below buds; bark not aromatic
.... hobblebush, *Viburnum alnifolium* (p. 361)

22′ Buds small, not stalked, superposed (two or more buds together), forming a close group, brown-hairy; leaf scars U-shaped or horseshoe shaped, more than half surrounding the buds; bark aromatic if cut or bruised........ sweetshrub, *Calycanthus* (p. 179)

21′ Buds scaly with two or more scales visible:

23 Pith very large in relation to size of twig, more than half and usually at least two-thirds the diameter of twig:

24 Bundle scars five to seven; leaf scars large, fan shaped, usually meeting around twig; lenticels prominent...... elder, *Sambucus* (p. 366)

24′ Bundle scars three; leaf scars smaller, crescent shaped, usually connected by a line; lenticels not prominent:

25 Twigs with two (sometimes three or four) hairy ridges extending downward from nodes; bark tight, not peeling
......................bush-honeysuckle, *Diervilla* (p. 352)

25' Twigs without hairy ridges; bark loosening and peeling after the first year...................... wild hydrangea, *Hydrangea arborescens* (p. 188)

23' Pith seldom more than one-half the diameter of twig:

26 Bud scales from last year's buds persisting at base of young twigs .... honeysuckle, *Lonicera* (p. 353)

26' Bud scales from last year's buds not persisting:

27 Terminal bud present:

28 Terminal bud with one pair of scales visible, the scales as long as the bud (valvate):

29 Young twigs dull colored (light brown, reddish brown, or grayish brown), often scurfy or hairy; flower buds larger than leaf buds, swollen near base of bud, long pointed at tip; buds either (*a*) brown, reddish brown, or dark red and scurfy or hairy or (*b*) grayish brown to purplish gray and covered with fine, scaly, waxy bloom .... viburnum, *Viburnum* (p. 358)

29' Young twigs often brightly colored (green, red, or purple); flower buds, if present, not readily distinguished from leaf buds; buds red to green, hairy or glabrous, not scurfy or waxy:

30 Buds ovoid, the terminal bud 5–15 mm long, the lateral buds prominently stalked, not much flattened........... maple, *Acer* (p. 263)

30' Buds conical, the terminal bud mostly 3–5 mm long, the lateral buds flattened, not stalked...................... silky dogwood, *Cornus amomum* (p. 302)

28' Terminal bud with more than one pair of scales visible, the scales overlapping, the first pair shorter than bud:

31 Bud scales two or three pairs:

32 Young twigs bright green; terminal bud densely covered with whitish hairs, the scales purplish beneath the hairs; opposite leaf scars meeting around twig and prolonged upward (toward twig tip) into points .... boxelder, *Acer negundo* (p. 265)

32' Young twigs not bright green; terminal bud not densely covered with whitish hairs; opposite leaf scars not meeting or, if they meet, not prolonged upward into points:

33 Bud scales with a rough-pitted or granular surface; bundle scars numerous, close together or touching, forming U-shaped band .... ash, *Fraxinus* (p. 335)

33' Bud scales not as above; bundle scars three:

      34 Twigs and buds orange to red; lateral buds rounded or bluntly pointed, not appressed; flower buds, or many of them, in clusters . . . . maple, *Acer* (p. 263)

      34' Twigs light reddish brown to gray; buds greenish to reddish brown; lateral buds sharply pointed and appressed; flower buds not clustered . . . . viburnum, *Viburnum* (p. 358)

   31' Bud scales four or more pairs:

      35 Leaf scars large, broad, shield shaped or triangular; bundle scars five or more, in a V-shaped line or in three groups . . . . buckeye, *Aesculus* (p. 272)

      35' Leaf scars smaller, narrow, V-shaped or U-shaped; bundle scars typically three, rarely five . . . . maple, *Acer* (p. 263)

27' Terminal bud absent:

   36 Twigs slender to moderate; leaf scars small, half circular or triangular to broadly crescent shaped; bundle scars three to seven:

      37 Stipule scars present, separate, distinct; bark of older twigs striped with white or green, not peeling . . . . bladdernut, *Staphylea trifolia* (p. 263)

      37' Stipule scars lacking; bark of older twigs loose and peeling, not striped . . . . hairy mock-orange, *Philadelphus hirsutus* (p. 187)

   36' Twigs stout; leaf scars large, vertically elliptical to nearly circular; bundle scars numerous in open or closed ellipse or ring:

      38 Pith continuous; leaf scars usually three at a node, two large scars and one smaller one; buds solitary . . . . catalpa, *Catalpa* (p. 346)

      38' Pith chambered or twigs sometimes hollow; leaf scars regularly two at a node; buds usually superposed . . . . royal paulownia, *Paulownia tomentosa* (p. 343)

Key IX. Deciduous Plants with Alternate Leaf Scars

(Leaf scars are sometimes crowded toward the tip of well-developed twigs or on alternate spur twigs.)

Most of the woody plants described in this book—representing about eighty-seven genera—have deciduous leaves and alternate leaf scars. Of this large number, some have winter characteristics that are easily overlooked or misinterpreted. The following key is arranged so that many of these woody plants can be identified even though not all the distinguishing traits have been observed. For example, if the naked catkins of eastern hophornbeam (*Ostrya virginiana*) are lacking or difficult to find, that species may also be keyed out under the heading "Bundle scars three or more." If the clustered buds at the tip of twigs in sweetleaf (*Symplocos tinctoria*) are not detected, that species may be found again under "Terminal bud present, at least twice as large as lateral buds." Not all the plants are repeated in this manner, but a number of them are—to help the beginner over some rough spots on the trail.

Index to Main Divisions of the Key

Read downward until you come to the first appropriate heading, then follow the key under number and page. **The main divisions of the key are in boldface type.**

Buds naked, or with leaflike scales, or the scales obscured by a dense covering of hair, wool, or scaly particles	60	70
Terminal bud three or four or more times as long as width of bud at base	73	71
Catkins or catkinlike flower buds visible in winter	80	72
Terminal bud present, at least twice as large as lateral buds	86	73
Buds superposed, one or two smaller buds at base of a larger bud	95	74
Flower buds distinguished from leaf buds by their different shape, larger size, or position on plant	101	75
Bundle scars three or more	109	76
Bundle scar one	109'	79

1 **Plants climbing, trailing, or sprawling;** stems needing support:
  2 Fruits grayish white or ivory colored, berrylike, about 5 mm across, borne in clusters in axils of leaf scars, often persistent into winter; *poisonous to touch*. . . . . poison-ivy, *Toxicodendron radicans* (p. 252)
  2' Fruits not as above or absent:
    3 Aerial roots present; terminal bud much larger than laterals, naked, slender, covered with light brown or light reddish brown hairs; *poisonous to touch* . . . . poison-ivy, *Toxicodendron radicans* (p. 252)
    3' Aerial rootlets absent; terminal bud lacking:
      4 Stems with tendrils:
        5 Tendrils attached to persistent base of leaf petioles; few to many prickles usually present; stems green, and apparently without pith . . . . greenbrier, catbrier, *Smilax* (p. 104)
        5' Tendrils attached to stem, opposite the leaf scars; prickles absent; stems not green; pith evident:
          6 Tendrils with three or more branches and with adhesive disks at tips; pith white or light green . . . . Virginia creeper, *Parthenocissus quinquefolia* (p. 284)
          6' Tendrils simple or merely forked, and without adhesive disks:
            7 Stems with brown pith:
              8 Pith continuous, without woody partition at each node; bark close and marked with conspicuous lenticels; tendrils simple . . . . muscadine, *Vitis rotundifolia* (p. 279)
              8' Pith interrupted by a woody partition at each node; bark loosening and peeling, the lenticels then not visible; tendrils forked. . . . . .grape, *Vitis* (p. 278)
            7' Stems with white pith, lacking a woody partition at each node but pith eventually separating into thin plates that begin

at outside and extend inward; bark with lenticels, not
loosening in thin strips. . . . . ampelopsis, *Ampelopsis* (p. 283)

4′ Stems without tendrils:

   9 Leaves deciduous above base of petiole, a part of old petiole
remaining, torn and shriveled, covering the leaf scar; stems often
prickly or bristly or both. . . . . . . blackberry, dewberry, *Rubus* (p. 212)

   9′ Leaves deciduous at base of petiole:

      10 Stems prickly; leaf scars very narrow, like a line, reaching
about halfway around stem . . . . .rose, *Rosa* (p. 219)

      10′ Stems not prickly:

         11 Leaf scars U-shaped; bundle scars three; buds superposed
on a hairy pad, surrounded on three sides by the leaf scar;
stems green . . . . Dutchman's-pipe,
Aristolochia durior (p. 166)

         11′ Leaf scars not U-shaped; bundle scars one to three or
more, often indistinct:

           12 Buds superposed, the upper bud in a depression a short
distance above leaf scar, the lower bud more or less
buried under leaf scar; leaf scars large, nearly or quite
as broad as stem, elliptical to circular, with raised
margin and concave center; bundle scars three to
seven:

              13 Young stems minutely hairy to glabrous; pith
white; leaf scars circular, commonly split at top
and showing one or two small buds beneath
. . . . .moonseed, *Menispermum canadense* (p. 172)

              13′ Young stems hairy; pith greenish; leaf scars half
circular with a concave upper margin
. . . . coralbeads, *Cocculus carolinus* (p. 172)

           12′ Buds solitary; leaf scars smaller, half circular, half
elliptical, or broadly crescent shaped; bundle scar one,
or several scars crowded together and appearing as
one:

              14 Buds ovoid to oblong, pointed, closely pressed to
stem, enclosed by two outer scales; leaf scars often
with a knobby growth at each side
. . . . wisteria, *Wisteria* (p. 238)

              14′ Buds rounded, not closely pressed to stem; leaf
scars without knobby growths:

                 15 Buds glabrous, projecting outward at right
angle to stem; bud scales several, the outer ones
keeled and pointed; pith solid
. . . . bittersweet, *Celastrus* (p. 260)

15' Buds densely hairy, the scales indistinct, and not projecting outward at right angle to stem; stems often three-sided or three-angled, with strong odor when bruised; pith hollow
....bitter nightshade, *Solanum dulcamara* (p. 342)

1' Plants erect or ascending; stems not needing support:
  16 **Plants with thorns, prickles, or stiff bristles:**
    17 Stems with thorns:
      18 Thorns in pairs at the leaf scars:
        19 Buds sunken under leaf scar and partly or entirely concealed; terminal bud absent; broken twigs with no strong odor
.... black locust, *Robinia pseudoacacia* (p. 242)
        19' Buds clearly visible, reddish brown, hairy; terminal bud present; broken twigs with strong lemon-peel odor
.... prickly-ash, *Zanthoxylum americanum* (p. 244)
      18' Thorns solitary (but they may be branched):
        20 Thorns attached just below lateral buds and often below short spurs bearing clusters of persistent leaf bases; thorns simple or three-branched; inner bark and wood yellow
...................................barberry, *Berberis* (p. 170)
        20' Thorns not attached as above, but occurring just above a leaf scar or terminating a branch:
          21 Twigs and buds covered with minute, silvery and/or yellowish brown, scaly particles; leaf scars half circular; bundle scar one .... autumn-olive, *Elaeagnus umbellata* (p. 294)
          21' Twigs and buds without such scaly particles; bundle scars more than one:
            22 Thorns regularly branched, very sharp pointed, borne on twigs, branches, and trunk; twigs conspicuously zigzag, swollen or knobby at nodes; buds nearly concealed under leaf scar or in bark above leaf scar
.... honeylocust, *Gleditsia triacanthos* (p. 232)
            22' Thorns single (occasionally branched in *Crataegus*); buds evident, not concealed:
              23 Thorns often bearing buds or leaf scars, or both:
                24 Terminal bud present; leaf scars linear or U-shaped:
                  25 Twigs mostly slender to moderate; buds and twigs grayish woolly, at least near tips, or if glabrous then bright red to dark red; lateral buds blunt to somewhat pointed, but not sharp pointed, and

usually closely pressed to twig
................crab apple, *Malus* (p. 208)
    25′ Twigs stout; buds and twigs glabrous or
    nearly so, brown to yellowish brown;
    lateral buds usually sharp pointed and
    slanting away from twig.........pear, *Pyrus*
    communis (p. 211)
  24′ Terminal bud absent; leaf scars half circular
  or elliptical; lateral buds brownish
......................plum, *Prunus* (p. 223)
23′ Thorns not or rarely bearing buds or leaf scars:
  26 Buds red or reddish, shiny, globe shaped;
  terminal bud present; sap not milky
................hawthorn, *Crataegus* (p. 204)
  26′ Buds light brown, flattened, partly concealed
  in bark of twig; terminal bud absent; sap
  milky (most easily seen in late winter, in cut
  section of twig) ....Osage-orange,
  Maclura pomifera (p. 161)
17′ Stems without thorns; prickles or stiff bristles, or both, present:
  27 Leaves deciduous 2–4 mm or more above base of petiole, leaving a
  remnant of petiole attached to stem, torn and stumplike, covering the
  true leaf scar; plants consisting of canes with soft wood and large pith;
  erect, arching, or trailing...............blackberry, dewberry, raspberry,
  Rubus (p. 212)
27′ Leaves deciduous at base of petiole:
  28 Buds hairy, without scales, partly or mostly buried under the leaf
  scar, only their tips showing; twigs set with stiff, bristly hairs
  ....bristly locust, *Robinia hispida* (p. 240)
  28′ Buds not buried under the leaf scar:
    29 Twigs stout; leaf scars very large, U-shaped, extending halfway
    or more around twig; bundle scars seven or more, in a single
    row................. devils-walkingstick, *Aralia spinosa* (p. 296)
    29′ Twigs slender; leaf scars linear to crescent shaped; bundle scars
    three:
      30 Bark loosening and shreddy (outer bark shed early); twigs
      prominently ridged downward from ends of leaf scars; pith
      with spongelike cavities, at least when dry; stems brown to
      gray............................. gooseberry, *Ribes* (p. 189)
      30′ Bark tight, not shreddy; twigs not ridged as above; pith
      solid; stems often highly colored, green or red...... rose, *Rosa*
      (p. 219)

16' Plants without thorns, prickles, or stiff bristles:
  31 **Twigs nearly or completely encircled by stipule scars at the nodes:**
    32 Leaf scars surrounding the buds; bud scale single, caplike; terminal
      bud absent .... sycamore, *Platanus occidentalis* (p. 196)
    32' Leaf scars not surrounding the buds; terminal bud present:
      33 Visible bud scales ten or more; stipule scars nearly meeting
        around twig; mature bark smooth, dark gray to almost white;
        twigs not aromatic .... American beech, *Fagus*
                                          grandifolia (p. 140)
      33' Visible bud scales only one or two; stipule scars extending all
        the way around twig; mature bark scaly or ridged and
        furrowed; twigs aromatic if cut or broken:
        34 Terminal bud apparently covered by a single scale (actually
          consisting of two united stipules) and either (*a*) ovoid and
          whitish-hairy or (*b*) long-conical, glabrous, purplish, often
          glaucous .... magnolia, *Magnolia* (p. 174)
        34' Terminal bud covered by two valvate scales, flattened and
          blunt, glabrous, green, purplish, or brownish red
          .... yellow-poplar, *Liriodendron tulipifera* (p. 177)
  31' Twigs without encircling stipule scars:
    35 **Buds crowded or clustered near tip of well-developed twigs** (twigs
      showing progressive shortening of internodes toward twig tip,
      causing gradual crowding of buds, should not be continued under
      this heading):
      36 Buds at twig tip small, partly submerged, indistinct, grayish-
        woolly with branched hairs (visible under magnifying lens);
        lateral buds partly or entirely sunken in tissue above leaf scars;
        threadlike or hairlike stipules often persistent; twigs commonly
        marked by circular scars indicating place of attachment left by
        fallen flowers .... rose-of-Sharon, *Hibiscus syriacus* (p. 289)
      36' Buds, at least terminal bud, prominently and fully exposed;
        stipules, if present, not threadlike or hairlike; circular scars left
        by fallen flowers lacking:
        37 Terminal bud grayish- or yellowish-hairy, appearing naked;
          bundle scar one, projecting out of triangular leaf scar; bark
          conspicuously loosening and curling; twigs commonly
          branching in first season, the branches much exceeding the
          terminal twig in length .... mountain pepperbush,
                                   Clethra acuminata (p. 303)
        37' Terminal bud (or buds in terminal cluster) scaly with two or
          more scales visible; twigs not branching in first season or, if
          branching, the lateral twigs not exceeding the terminal twig
          in length:

38 Leaf scars extending halfway or more around stem; roots and inner bark yellow ..... yellowroot, *Xanthorhiza simplicissima* (p. 168)

38′ Leaf scars not extending halfway or more around stem:

39 First (lowest) scale of lateral buds centered directly above leaf scar; bark of twigs very bitter .... cottonwood, poplar, *Populus* (p. 112)

39′ First (lowest) scale of lateral buds not centered directly above leaf scar, or position of scale difficult to determine:

40 Pith of twigs chambered; bundle scar one, crescent shaped .... sweetleaf, *Symplocos tinctoria* (p. 332)

40′ Pith of twigs not chambered:

41 Bundle scar one; fruit a capsule, often present in winter:

42 Older twigs usually with shredding bark; capsule 4–7 mm long .... minnie-bush, *Menziesia pilosa* (p. 312)

42′ Older twigs usually with tight bark; capsule more than 7 mm long .... azalea, *Rhododendron* (p. 304)

41′ Bundle scars three or more; fruit not a capsule:

43 Bundle scars three; buds with about six overlapping scales; crushed twigs emitting strong odor of bitter almonds; bark on older branches banded crosswise with elongated lenticels .... cherry, *Prunus* (p. 223)

43′ Bundle scars numerous, scattered; buds with many overlapping scales; crushed twigs with no strong odor; bark not banded with elongated lenticels .... oak, *Quercus* (p. 142)

35′ Buds not crowded or clustered near tip of well-developed twigs:

44 **Leaf scars nearly surrounding the buds:**

45 Twigs bristly with coarse, stiff, rusty brown hairs and usually bearing a pair of weak thorns at each node .... bristly locust, *Robinia hispida* (p. 240)

45′ Twigs not bristly; thorns absent:

46 Leaf scars forming a nearly closed ring around buds, open only at top:

47 Twigs very stout and stiff; sap milky; bundle scars
numerous, scattered or in groups; buds solitary
.... sumac, *Rhus* (p. 248)

47′ Twigs slender to moderate; sap not milky; bundle
scars usually five:

     48 Buds solitary; twigs swollen at nodes, flexible
but with very tough bark, difficult to break by
hand .... leatherwood, *Dirca palustris* (p. 294)

     48′ Buds superposed, two to four together, forming
a cone-shaped group; bundle scars projecting
above surface of leaf scars; bark of trunk smooth
and gray, similar to that of beech
.... yellowwood, *Cladrastis kentukea* (p. 234)

46′ Leaf scars U-shaped, on three sides of bud only:

     49 Bundle scars three; buds silvery or white; sap not
resinous; leaf scars often torn or cracked by buds;
broken twigs with strong sour odor, suggesting
strong lemon peel .... hoptree, *Ptelea
trifoliata* (p. 244)

     49′ Bundle scars five to nine; buds yellowish brown; sap
resinous; leaf scars usually not disturbed by buds;
orange lenticels prominent; twigs without strong
odor .... shining sumac, *Rhus copallina* (p. 250)

44′ Leaf scars below the buds, not nearly surrounding them:

     50 **Leaves deciduous 2–4 mm above base of petiole, leaving basal part
persistent on stem:**

         51 Persistent base of petioles torn and shriveled; leaf scars usually not
distinguishable; twigs brierlike, often prickly
.... blackberry, dewberry, raspberry, *Rubus* (p. 212)

         51′ Persistent base of petioles bearing evident leaf scars; twigs not
brierlike, never prickly:

             52 Leaf scars circular or nearly so; twigs slender; catkinlike flower
buds often exposed; bushy shrub .... fragrant sumac,
Rhus aromatica (p. 249)

             52′ Leaf scars crescent shaped to broadly U-shaped; twigs moderate
to stout; catkinlike flower buds absent; tree or large shrub
.... American mountain-ash, *Sorbus americana* (p. 205)

     50′ Leaves deciduous at base of petiole (although leaf scars may be raised):

         53 **Buds more or less buried under the leaf scar, or concealed in bark
or between twig and petiole base, or so flattened against twig as to
appear partly concealed** (see also *Hibiscus*):

             54 Leaf scars almost circular, much raised on persistent base of
petiole; leaf buds small, yellow-hairy, concealed between twig

and petiole base; catkinlike flower buds often present; bark with
fragrant odor if cut or bruised .... fragrant sumac,
Rhus aromatica (p. 249)

54′ Leaf scars not as above; catkinlike flower buds never present:

55 Buds, or most of them, buried under the leaf scar and
surrounded or nearly surrounded by it; buds two or more at
a node (superposed); leaf scars torn or distorted by growth
of buds in spring:

56 Stipules present, weak, hairlike or bristlelike, or leaving
small scars; all buds at each node sunken under the leaf
scar.......................... locust, *Robinia* (p. 238)

56′ Stipules and stipule scars lacking; one or more buds at
each node in bark of twig above leaf scar:

57 Twigs reddish brown to greenish red, prominently
enlarged at nodes; uppermost bud at each node
appearing as a tiny knob pushing through the bark;
plants usually armed with stout thorns, many of
them branched .... honeylocust, *Gleditsia
triacanthos* (p. 232)

57′ Twigs some shade of green, not or only slightly
enlarged at nodes; uppermost bud at each node
small but fully exposed, two- to several-scaled,
glabrous; lower buds partly or entirely sunken in
bark or concealed under leaf scar; leaf scars with a
smooth, shiny border; plants never armed with
thorns .... silktree, *Albizia julibrissin* (p. 230)

55′ Buds not buried under the leaf scar but more or less sunken
in bark, or flattened against twig, above leaf scar; leaf scars
not or very little disturbed by growth of buds in spring:

58 Twigs stout and blunt; leaf scars large and conspicuous;
bundle scars three to many; pith variously colored, not
white:

59 Bundle scars numerous, arranged in a row near edge
of leaf scar; buds solitary; twigs yellowish brown
tinged with green; pith tan to chocolate brown;
broken twigs with disagreeable odor
.... ailanthus, *Ailanthus altissima* (p. 245)

59′ Bundle scars three to five; buds superposed, sunken
in hairy depressions above leaf scar; twigs grayish to
brown, covered with gray skin; pith pink or red;
twigs without disagreeable odor
... Kentucky coffeetree, *Gymnocladus dioicus* (p. 232)

58′ Twigs slender to moderate, dark red to greenish red; leaf

scars smaller, shield shaped; bundle scar one, crescent shaped or V-shaped; pith white
.... sourwood, *Oxydendrum arboreum* (p. 320)

53′ Buds evident on well-developed growth, not buried or concealed or flattened (some buds may be clearly visible but appressed to twig):

  60 **Buds naked, or with leaflike scales, or the scales obscured by a dense covering of hair, wool, or scaly particles:**

    61 Buds and twigs with a scurfy coating of minute silvery and/or golden brown scales; bundle scar one; twigs often ending in a thorn
.... autumn-olive, *Elaeagnus umbellata* (p. 294)

    61′ Buds and twigs not coated with silvery or golden brown scales:

      62 Stems very slender, straight, erect, unbranched or sparingly branched, covered with rusty wool; buds very small, rusty brown–woolly; leaf scars indistinct, much raised, bordered by torn edges of petiole base; buds, leaf scars difficult to see clearly without magnifying lens .... steeplebush spiraea,
Spiraea tomentosa (p. 199)

      62′ Stems not as above in all respects; buds and leaf scars larger:

        63 Buds, at least some of them, stalked:

          64 Buds either bright yellow or dark brown to red; stipule scars absent; terminal bud not stalked; lateral buds often superposed:

            65 Terminal bud yellow, scurfy, with four to six scales, the latter valvate in pairs but bud appearing naked; leaf scars large, heart shaped or three-lobed, with numerous bundle scars; lateral buds with valvate scales, stalked .... bitternut hickory,
Carya cordiformis (p. 122)

            65′ Terminal bud dark red or brown, hairy, flattened, naked; leaf scars smaller, broadly crescent shaped, with five or rarely seven bundle scars; lateral flower buds rounded, stalked .... pawpaw, *Asimina triloba* (p. 180)

          64′ Buds not bright yellow and not dark brown to red; stipule scars present; terminal bud stalked; lateral buds not superposed but sometimes collateral:

            66 Terminal bud yellowish brown, distinctly flattened, sickle shaped or crescent shaped, naked except for two narrow scalelike stipules, these falling off early and leaving short scars; flower buds axillary, flowers blooming in late autumn to early winter
.... witch-hazel, *Hamamelis virginiana* (p. 192)

            66′ Terminal bud medium brown, somewhat compressed but not flattened as above; outer scales two, these deciduous early or sometimes broken off, exposing lighter colored, densely hairy bud beneath; flower buds terminal, with a long, narrow, keeled scale, enlarged at tip, visible along one side of bud; flowers blooming in spring . . . . large fothergilla, *Fothergilla major* (p. 193)

63′ Buds not stalked:

67 Bundle scar one, protruding above surface of leaf scar:

   68 Twigs densely woolly with branched hairs; terminal bud not flattened, yellowish-hairy; bark of older twigs peeling and curling; twigs commonly branching in first year, the branches surpassing the terminal twig in length . . . . mountain pepperbush, *Clethra acuminata* (p. 303)

   68′ Twigs glabrous or with only scattered hairs; terminal bud flattened and twisted, silvery-silky; twigs not or rarely branching in first year . . . . mountain camelia, *Stewartia ovata* (p. 290)

67′ Bundle scars three or more:

   69 Leaf scars large, more or less three-lobed, with three groups of bundle scars, each group usually U-shaped:

      70 Pith of twigs light brown or dark brown, chambered, the chambers empty; terminal bud present; lateral buds often superposed . . . . walnut, *Juglans* (p. 117)

      70′ Pith of twigs white, continuous; terminal bud absent; lateral buds solitary . . . . chinaberry, *Melia azedarach* (p. 247)

   69′ Leaf scars smaller, crescent shaped to elliptical; bundle scars three; pith continuous; lateral buds solitary:

      71 Terminal bud present, larger than lateral buds:

         72 Terminal bud gray-woolly, with about four overlapping scales; stipule scars absent; lateral buds somewhat flattened against twig; spur twigs common . . . . apple, *Malus sylvestris* (p. 210)

         72′ Terminal bud light brown or golden brown to dark brown, hairy, naked (without typical scales); stipule scars present; spur twigs few or none . . . . buckthorn, *Rhamnus* (p. 275)

      71′ Terminal bud absent; lateral buds covered by a single caplike scale . . . . willow, *Salix* (p. 108)

60′ Buds scaly, with one or more scales clearly visible:

73 **Terminal bud three or four or more times as long as width of bud at base:**

   74 Buds spindle shaped, with ten or more scales exposed; stipule scars extending almost around twig; lateral buds widely divergent . . . . American beech, *Fagus grandifolia* (p. 140)

74' Buds not spindle shaped, with seven or fewer scales exposed; stipule scars shorter than above or absent:

75 Bundle scars about ten; leaf scars more than half encircling the stem; inner bark yellow .... yellowroot, *Xanthorhiza simplicissima* (p. 168)

75' Bundle scars three; leaf scars not more than half encircling the stem; inner bark not yellow:

76 Twigs stout; leaf scars broad, three-lobed or triangular; first (lowermost) scale of lateral buds large and centered directly above leaf scar.......... eastern cottonwood, balm-of-Gilead, *Populus* (p. 112)

76' Twigs slender; leaf scars long and narrow, crescent shaped or U-shaped; first (lowermost) scale of lateral buds not directly above leaf scar:

77 Older twigs with many short spurs, each bearing crowded leaf scars and ending in a terminal bud:

78 Twigs with prominent ridges extending downward from ends of leaf scars; stipule scars absent; plants often with prickles .... gooseberry, *Ribes* (p. 189)

78' Twigs without prominent ridges extending downward from ends of leaf scars; stipule scars present; plants without prickles.......... birch, *Betula* (p. 129)

77' Older twigs without many short spurs as above:

79 Terminal bud distinctly flattened, either dark red or densely woolly; next-to-lowest scale of terminal bud one-half or more as long as bud .... chokeberry, *Aronia* (p. 202)

79' Terminal bud not flattened, delicately colored, tinged with pink or green (may be reddish to purplish on one side); next-to-lowest scale of terminal bud less than half as long as bud ....serviceberry, *Amelanchier* (p. 206)

73' Terminal bud lacking or, if present, less than three or four times as long as width of bud at base:

80 **Catkins or catkinlike flower buds visible in winter:**

81 Fruit in conelike structures, woody, persisting through winter after seeds have fallen, conspicuous, often visible from a distance; leaf buds long stalked (merely short stalked or stalkless in *A. crispa*); stipule scars present .... alder, *Alnus* (p. 133)

81' Fruit not as above:

82 Bundle scar one; flower buds in one-sided, elongated

clusters (racemes); twigs often reddish brown above, green
beneath; stipule scars lacking. . . . . sweetbells, *Leucothoe*
(p. 317)

82′ Bundle scars three or more; stipule scars present:

   83 Spur twigs numerous on older growth, each bearing
crowded leaf scars and one bud only, a terminal bud;
buds stalkless but commonly appearing stalked at ends
of short spurs; terminal bud absent on long twigs;
lenticels conspicuous, elongated into transverse lines or
bands on branches and trunk . . . . birch, *Betula* (p. 129)

   83′ Spur twigs absent or at least not as above; lenticels not
elongated into transverse lines or bands:

      84 Twigs dotted with yellow resin droplets (visible
under magnifying lens), aromatic if cut or broken;
bundle scars three, distinct, the middle one largest;
leaves fernlike, deciduous but often persisting in
winter . . . . . . sweet-fern, *Comptonia peregrina* (p. 117)

      84′ Twigs without resin droplets, not aromatic; leaves
not fernlike:

         85 Bundle scars three; bud scales (under magnifying
lens) marked with fine parallel grooves; bark of
trunk in narrow vertical strips loose at the ends
. . . . eastern hophornbeam,
Ostrya virginiana (p. 137)

         85′ Bundle scars usually more than three, scattered
and indistinct; bud scales not marked as above;
bark smooth, thin, brown or gray
. . . . hazelnut, *Corylus* (p. 137)

80′ Catkins or catkinlike flower buds absent in winter:

 86 **Terminal bud present, at least twice as large as lateral buds:**

   87 Leaf scars large and prominent, triangular, shield shaped, or heart
shaped; bundle scars numerous (*caution: poison-sumac is in this
group*):

      88 Pith chambered, the chambers hollow; bundle scars in three
clearly defined, U-shaped groups . . . . walnut, *Juglans* (p. 117)

      88′ Pith continuous; bundle scars in three or more groups or
scattered:

         89 Terminal bud sharp-conical, purplish; lateral buds solitary;
pith large, circular in cross section; *poisonous to touch;
handle with care* . . . . poison-sumac, *Toxicodendron
vernix* (p. 254)

         89′ Terminal bud ovoid to nearly globose, not purplish; lateral
buds often superposed; pith small, angled in cross section
. . . . hickory, *Carya* (p. 120)

87′ Leaf scars smaller, half circular to crescent shaped or linear; bundle scars one to about five:

  90 First-year growth often branched, the terminal twig surpassed in length by the lateral twig just below it:

    91 Lateral buds evident; terminal bud with four or five scales exposed; bark and buds with spicy odor and taste; bundle scar one ....sassafras, *Sassafras albidum* (p. 182)

    91′ Lateral buds concealed behind persistent base of leaf petioles; terminal bud with two or three scales exposed; bark and buds with unpleasant odor and bitter taste; bundle scars three .... alternate-leaf dogwood, *Cornus alternifolia* (p. 299)

  90′ First-year growth rarely branched; if branched, not as above:

    92 Bundle scar one, appearing as a transverse line or crescent shaped; pith chambered, the chambers empty; large, globe-shaped flower buds often present ....sweetleaf, *Symplocos tinctoria* (p. 332)

    92′ Bundle scars three or more; pith not chambered:

      93 Leaf scars raised on dark red ledges (remains of petiole base), these later deciduous, exposing the true leaf scars; bundle scars usually five; terminal bud large, 9–18 mm long, dark red or purplish red, gummy ....American mountain-ash, *Sorbus americana* (p. 205)

      93′ Leaf scars not raised as above; bundle scars three; terminal bud smaller, 2–12 mm long:

        94 Buds and twigs more or less woolly, or if glabrous then terminal bud only 2–5 mm long; older twigs without corky ridges; terminal bud sometimes lacking on long growths .... apple, *Malus* (p. 208)

        94′ Buds and twigs glabrous; terminal bud 6–12 mm long, glossy, fragrant if crushed; older twigs often with corky ridges of bark .....sweetgum, *Liquidambar styraciflua* (p. 192)

86′ Terminal bud absent, or if present not much if any larger than lateral buds:

  95 Buds superposed, one or two smaller buds at base of a larger bud:

    96 Pith chambered, the chambers either empty or filled with loose tissue:

      97 Bark of twigs becoming silky-shreddy by the second or third year; young trunks and branches white striped; bundle scar one ....Carolina silverbell, *Halesia carolina* (p. 333)

      97′ Bark of twigs remaining tight; young trunks and branches not white striped; bundle scars three:

98 Twigs brown, reddish brown, or greenish brown; bundle
    scars usually sunken below surface of leaf scar; lateral buds
    often short stalked . . . . . . black tupelo, *Nyssa sylvatica* (p. 298)

98′ Twigs green; bundle scars not sunken; lateral buds not
    stalked. . . . . sweet-spires, *Itea virginica* (p. 185)

96′ Pith not chambered:

  99 Terminal bud present; bundle scar one; broken bark not
    aromatic:

    100 Stipules present (under magnifying lens), very small,
      triangular, or leaving minute scars; pith continuous; buds
      with two to six scales exposed. . . . . holly, *Ilex* (p. 255)

    100′ Stipules and stipule scars lacking; pith spongy; buds with
      two scales exposed. . . . . stewartia, *Stewartia* (p. 289)

  99′ Terminal bud absent; bundle scars three; stipules slender,
    pointed, purplish, or their scars present; broken bark mildly
    aromatic . . . . indigobush, *Amorpha* (p. 236)

95′ Buds not superposed:

  **101 Flower buds distinguished from leaf buds by their different shape,
    larger size, or position on plant:**

  102 Flower buds stalked; bundle scars three:

    103 Flower buds with eight or more scales exposed, the latter
      progressively shorter toward base of bud, covering the stalk;
      leaf scars fringed with hairs on upper edge; flower buds
      often paired or clustered on larger twigs and branches;
      twigs reddish brown to purplish . . . . .redbud, *Cercis
      canadensis* (p. 230)

    103′ Flower buds with only a few scales exposed, the stalk
      clearly evident; twigs and buds with spicy odor and taste;
      flower buds often in pairs, one on each side of a leaf bud;
      twigs green to greenish brown. . . . . .spicebush, *Lindera
      benzoin* (p. 182)

  102′ Flower buds not stalked:

    104 Flower buds in long, narrow, dark red to greenish clusters
      (racemes), mostly 3 cm or more long, each bud in the axil of
      a scalelike bract; leaf buds very small, less than 2 mm long,
      roundish, widely divergent; bundle scar one
      . . . . sweetbells, *Leucothoe* (p. 317)

    104′ Flower buds not in long, narrow clusters as above, but
      readily separated from leaf buds by their larger size:

      105 Bundle scar one; stipule scars absent:

        106 Terminal bud present; pith chambered; flower
          buds large, globe shaped; leaf buds very small,
          conical . . . . . sweetleaf, *Symplocos tinctoria* (p. 332)

106′ Terminal bud absent; pith continuous:

  107 Flower buds sprinkled with yellow resinous droplets, at least near tip of bud (visible under magnifying lens); outer scales of leaf buds not prolonged into points .... huckleberry, *Gaylussacia* (p. 321)

  107′ Flower buds without such resinous droplets; outer scales of leaf buds prolonged into slender points ....blueberry, *Vaccinium* (p. 324)

105′ Bundle scars typically three; stipule scars present:

  108 Twigs slender to moderately stout; bundle scars sunken below surface of leaf scars, easily seen; bud scales in two vertical rows; mature bark with ridges and furrows.............elm, *Ulmus* (p. 155)

  108′ Twigs very slender; bundle scars not sunken, often indistinct; bud scales in four vertical rows; mature bark bluish gray, smooth and close fitting ....American hornbeam, *Carpinus caroliniana* (p. 135)

101′ Flower buds either not present or not distinguishable from leaf buds by their size, shape, or position on plant:

109 **Bundle scars three or more:**

  110 Buds with only one scale, covering bud like a hood or inverted bag (see also *Magnolia* and *Platanus*)....... willow, *Salix* (p. 108)

  110′ Buds with two or more scales showing:

    111 Visible bud scales two or three:

      112 Pith with a single green partition at each node; sap milky; buds sometimes opposite .... paper-mulberry, *Broussonetia papyrifera* (p. 161)

      112′ Pith without a green partition at each node; sap not milky; buds regularly alternate:

        113 Buds with usually two visible scales (sometimes tip of third scale showing); lenticels not elongated:

          114 Buds distinctly lopsided, the outermost scale bulging; twigs usually zigzag and, if carefully sliced crosswise, showing a ring of triangles in the bark; pith circular or nearly so in cross section .... basswood, *Tilia* (p. 286)

          114′ Buds not distinctly lopsided; twigs nearly straight and without such triangles in bark;

pith five-sided in cross section
...... chestnut, chinkapin, *Castanea* (p. 140)

113' Buds with three (rarely four) visible scales, the
lower pair forming a V-shaped angle over the
leaf scar; lenticels elongated to horizontal stripes
on older twigs and branches .... birch, *Betula*
(p. 129)

111' Visible bud scales four or more:

115 Terminal bud present:

116 First (lowermost) scale of lateral buds centered directly
above the leaf scar .... cottonwood, poplar, *Populus* (p. 112)

116' First (lowermost) scale of lateral buds not centered directly
above leaf scar:

117 Bundle scars five, the upper pair very small, the
middle one largest and prominent; bark shredding or
peeling in long strips; twigs strongly ridged
downward from nodes; stipule scars present
.... ninebark, *Physocarpus opulifolius* (p. 198)

117' Bundle scars three; bark not shredding or peeling as
above:

118 Twigs, when bruised, emitting odor of bitter
almonds; older bark prominently marked with
elongated lenticels; stipule scars present
.... cherry, peach, *Prunus* (p. 223)

118' Twigs without bitter almond odor; stipule scars
absent:

119 Pith solid but chambered, with partitions
of firmer tissue at intervals; leaf scars broad
and rounded, half circular to somewhat
heart shaped, the bundle scars usually
depressed below surface of leaf scar
.... black tupelo, *Nyssa sylvatica* (p. 298)

119' Pith not chambered; leaf scars long and
narrow, crescent shaped to U-shaped, the
bundle scars not depressed:

120 Twigs rather stout and heavy; spur
twigs numerous, often thorn tipped;
terminal bud reddish brown, not
flattened ....pear, *Pyrus communis*
(p. 211)

120' Twigs slender to moderate; spur twigs
few or none; terminal bud pinkish to
dark red or purplish red, strongly
flattened ... chokeberry, *Aronia* (p. 202)

115′ Terminal bud absent:
  121 Bundle scars three:
    122 Buds large, greenish, widely divergent, solitary or in irregular groups of two or more, each bud with many scales showing; first-year twigs commonly self-pruned, leaving conspicuous branch scars with many bundle scars arranged in an ellipse . . . . .buffalo nut, *Pyrularia pubera* (p. 165)
    122′ Buds without the foregoing combination of characters; first-year twigs normally not self-pruned:
      123 Pith finely chambered (partitions close together), at least near nodes; buds flattened, closely appressed to twig . . . . .hackberry, *Celtis* (p. 158)
      123′ Pith not chambered:
        124 Leaf scars more than two-ranked; older bark strongly banded crosswise with elongated lenticels; spur twigs often thorn tipped. . . . . .plum, *Prunus* (p. 223)
        124′ Leaf scars two-ranked; bark not banded crosswise with elongated lenticels; spur twigs, if present, not thorn tipped:
          125 Bud scales in two vertical rows; bundle scars three or in three groups, sunken below surface of leaf scar; buds often tipped to one side of leaf scar. . . . . elm, *Ulmus* (p. 155)
          125′ Bud scales not in two vertical rows; bundle scars not sunken:
            126 Buds 3–7 mm long, not angled; bud scales marked vertically with fine parallel grooves (under magnifying lens); bark of trunk brownish gray, in narrow vertical strips . . . . .eastern hophornbeam, *Ostrya virginiana* (p. 137)
            126′ Buds 2–4 mm long, commonly four-angled; bud scales not marked with fine grooves; bark of trunk bluish gray, smooth and tight, appearing "muscular" . . . . .American hornbeam, *Carpinus caroliniana* (p. 135)
  121′ Bundle scars more than three:
    127 Twigs stout (very stout on vigorous growth); leaf scars large and prominent, heart shaped to shield shaped; bundle scars many, in a row near edge of leaf scar; buds small, 2 mm long or less, dome shaped, with two or sometimes four scales exposed . . . . ailanthus, *Ailanthus altissima* (p. 245)

127′ Twigs slender to moderate; leaf scars smaller, half circular to nearly circular or triangular; bundle scars not in a row near edge of leaf scar; buds 2–5 mm long, with four to seven scales exposed:

    128 Sap watery-milky; bundle scars numerous, in a closed ring or scattered, projecting from surface of concave leaf scar; bud scales in two rows . . . . mulberry, *Morus* (p. 162)

    128′ Sap not watery-milky; bundle scars several (occasionally three), scattered and indistinct; bud scales not in two rows . . . . hazelnut, *Corylus* (p. 137)

109′ **Bundle scar one,** or several scars crowded together and appearing as one (may be spread out in a straight or curved line):

    129 Leaf scars very small, much raised above surface of twig, often bordered by a pale area or by torn edges of petiole base; buds small, solitary, with several scales visible; buds, leaf scars difficult to examine without magnifying lens; stems usually low, very slender, sparingly branched or stiff and erect. . . . . spiraea, *Spiraea* (p. 199)

    129′ Leaf scars larger, not so much raised above twig surface, and not bordered as above:

        130 Stems with two kinds of scars: (*a*) circular branch scars, left by self-pruned twigs, with many bundle scars arranged in a ellipse, and (*b*) semicircular to shield shaped leaf scars with typically one bundle scar; branch scars above leaf scars; buds large, greenish, usually set at wide angle to twig, often irregularly clustered . . . . buffalo nut, *Pyrularia pubera* (p. 165)

        130′ Stems with only leaf scars; branch scars lacking:

            131 Stems and branches green, decidedly angled and grooved, almost winged at the angles, long and whiplike, often dying back from tips . . . . Scotch broom, *Cytisus scoparius* (p. 235)

            131′ Stems and branches not as above in all respects:

                132 Bud scales two, valvate; buds slender, 3–6 mm long, pointed, flattened, usually appressed to twig:

                    133 Twigs yellow to yellowish brown; buds pinkish red; leaf scars shield shaped; fruits small, globose capsules, persistent in winter . . . . male-berry, *Lyonia ligustrina* (p. 316)

                    133′ Twigs dark red to green; buds brownish red to green; leaf scars crescent shaped to half elliptical; fruits fleshy berries, rarely present in winter. . . . . bearberry, *Vaccinium erythrocarpum* (p. 327)

132′ Bud scales more than two, or if only two, (*a*) the
scales not valvate or (*b*) the buds not flattened and
appressed:

134 Bundle scars crescent shaped or U-shaped:

135 Bark of twigs becoming silky-shreddy by
second year; buds red or greenish red; bark
of young trunks and branches streaked
with chalky white.........Carolina silverbell,
Halesia carolina (p. 333)

135′ Bark of twigs not silky-shreddy:

136 Buds small, mostly less than 2 mm
long, appearing partly submerged in
bark, rounded-conical, reddish, with
four to six visible scales; twigs red to
green .... sourwood, *Oxydendrum
arboreum* (p. 320)

136′ Buds larger, 2–4 mm long, fully
exposed, ovoid, dark reddish brown
to purplish, with two visible scales,
one scale greatly overlapping the
other; twigs purplish to grayish
brown .... persimmon, *Diospyros
virginiana* (p. 331)

134′ Bundle scars not crescent shaped or U-shaped:

137 Twigs clustered at upper end of each year's growth
.... azalea, *Rhododendron* (p. 304)

137′ Twigs not clustered as above:

138 Plants low, scarcely woody; fruits dry capsules, in stalked
clusters, falling when mature, the saucer-shaped bases
persistent in winter, distinctive .... New Jersey tea,
Ceanothus americanus (p. 276)

138′ Plants definitely woody; fruits fleshy, without saucer-
shaped bases as above:

139 Stipule scars present, one at each end of a leaf scar, or
minute, triangular, pointed stipules often persistent
.... holly, *Ilex* (p. 255)

139′ Stipule scars and stipules lacking:

140 Buds small, mostly less than 2 mm long, short-
ovoid to nearly globose, set at wide angle to
twig.......... sparkleberry, deerberry, *Vaccinium*
(p. 324)

140′ Buds not as above:

141 Twigs finely speckled with warty dots (may

be obscured by hairs); buds with the lowest
scales extended into long, slender points
.... blueberry, *Vaccinium* (p. 324)

141′ Twigs not speckled with warty dots; bud
scales more or less pointed but not
extended into long, slender points
.... huckleberry, *Gaylussacia* (p. 321)

Shortcuts to Identifying Some Woody Plants

Some woody plants have one or more especially noticeable characters, such as color of twigs, shape of terminal buds, or presence of naked catkins, by which they can be quickly recognized. These plants have been placed in the following groups. Note that the groups do not cover *all* the woody plants in our area, but only those with unusual or particularly striking characteristics. The groups do not include members of the Pine Family and the Cypress Family, nor do they cover plants having evergreen leaves.

Most of the groupings are arranged roughly in the form of keys, and so you should read through the descriptions until you find one that fits the specimen in hand. If the plant cannot be found in one of these groups, you will need to use one of the longer keys.

I *Plants with Climbing Stems (Vines)*

1 Plants climbing by means of aerial rootlets:
  2 Leaves opposite:
    3 Leaves simple, entire or toothed..................southern decumaria
                                    (*Decumaria barbara*) (p. 185)
    3′ Leaves compound, with seven to fifteen coarsely toothed leaflets
      ....trumpet creeper (*Campsis radicans*) (p. 345)
  2′ Leaves alternate:
    4 Leaves simple, lobed, evergreen........English ivy (*Hedera helix*) (p. 295)
    4′ Leaves compound, with three leaflets, deciduous; *poisonous to touch*
      ....poison-ivy (*Toxicodendron radicans*) (p. 252)
1′ Plants climbing by means of tendrils:
  5 Tendrils attached to leaf petioles; stems green and usually prickly
    .... greenbrier, catbrier (*Smilax*) (p. 104)
  5′ Tendrils attached to stems, opposite the leaves; stems never prickly:
    6 Leaves simple:
      7 Pith brown.....................................grape (*Vitis*) (p. 278)
      7′ Pith white........ heartleaf ampelopsis (*Ampelopsis cordata*) (p. 284)
    6′ Leaves compound:
      8 Leaflets numerous, arranged pinnately................... pepper-vine
                           (*Ampelopsis arborea*) (p. 283)
      8′ Leaflets usually five, arranged palmately ....Virginia creeper
                       (*Parthenocissus quinquefolia*) (p. 284)
      8″ Leaflets usually two, with a branched tendril between them
      ....cross vine (*Bignonia capreolata*) (p. 345)

1″ Plants climbing by means of twining petioles; leaves opposite, compound
....clematis (*Clematis*) (p. 168)

1‴ Plants climbing by means of twining stems (or stems sometimes only leaning
    in *Solanum*):

   9 Leaves opposite, simple..........................honeysuckle (*Lonicera,*
                   especially *L. japonica*) (p. 355)

   9′ Leaves alternate:

      10 Leaves compound, with nine to fifteen leaflets .... wisteria (*Wisteria*)
                   (p. 238)

      10′ Leaves simple:

         11 Leaves toothed ....bittersweet (*Celastrus*) (p. 260)

         11′ Leaves not toothed:

            12 Leaves, at least some of them, deeply two-lobed near base
              ....bitter nightshade (*Solanum dulcamara*) (p. 342)

            12′ Leaves entire or variously lobed or angled, with the petiole
              attached to lower side of blade a short distance inside margin;
              stems finely ridged lengthwise........ moonseed (*Menispermum*
                   canadense) (p. 172)

            12″ Leaves entire, heart shaped to nearly circular in outline;
              stems enlarged at the nodes ....Dutchman's-pipe
                   (*Aristolochia durior*) (p. 166)

II *Plants with Milky Sap*
(best seen at the base of a broken petiole)

1 Leaves compound; leaflets eleven to thirty-one .... smooth and staghorn sumac
                   (*Rhus*) (p. 248)

1′ Leaves simple:

   2 Leaves entire; stems thorny ...... Osage-orange (*Maclura pomifera*) (p. 161)

   2′ Leaves toothed, either lobed or unlobed ....mulberry (*Morus*) (p. 162)
                 paper-mulberry (*Broussonetia papyrifera*) (p. 161)

III *Plants with Naked Catkins from Late Summer through All or Most of Winter*

1 Woody fruit "cones" from previous years persisting on plant ..... alder (*Alnus*)
(p. 133)
2 Older twigs with many short growths, each regularly bearing two leaves and only one bud, a terminal bud........................birch (*Betula*) (p. 129)
3 Leaves fernlike, aromatic; low shrub .... sweet-fern (*Comptonia peregrina*) (p. 117)
4 Leaves much longer than wide, similar to those of elm or birch; catkins often in threes .... eastern hophornbeam (*Ostrya virginiana*) (p. 137)
5 Leaves not much longer than wide, irregularly double toothed
.... hazelnut (*Corylus*) (p. 137)

IV *Terminal Buds Naked or at Least Appearing Naked*
(without typical scales)

1 Leaves compound, alternate:
  2 Terminal buds yellow; pith solid....... bitternut hickory (*Carya cordiformis*)
(p. 120)
  2' Terminal buds grayish to brownish; pith chambered .... walnut (*Juglans*)
(p. 117)
1' Leaves simple:
  3 Terminal buds long stalked:
    4 Leaves opposite, finely toothed, rounded or heart shaped at base
.... hobblebush (*Viburnum alnifolium*) (p. 361)
    4' Leaves alternate, coarsely round toothed, lopsided at base
..........................witch-hazel (*Hamamelis virginiana*) (p. 192)
  3' Terminal buds not stalked:
    5 Buds dark red...................... pawpaw (*Asimina triloba*) (p. 180)
    5' Buds light brown .... Carolina buckthorn (*Rhamnus caroliniana*)
(p. 275)

V   *Buds Long, Very Slender, and Sharply Pointed*

1 Leaves clustered at the top of the stem, compound, the leaflets sharply and irregularly toothed or lobed; inner bark and roots deep yellow; low, few-branched shrub . . . . yellowroot (*Xanthorhiza simplicissima*) (p. 168)

2 Leaves with a row of dark glands along midrib on upper side; terminal buds markedly flattened . . . . chokeberry (*Aronia*) (p. 202)

3 Buds resinous and fragrant; bark of twigs very bitter . . . . cottonwood (*Populus*) (p. 112)

4 Buds pinkish, greenish, or reddish, with five or six scales showing; lateral veins of leaves curving and branching before reaching margin . . . . downy serviceberry (*Amelanchier arborea*) (p. 207)

5 Buds reddish brown, with ten or more scales showing; veins of leaves running straight and parallel from the midrib to the margin. . . . American beech (*Fagus grandifolia*) (p. 140)

VI   *Stems Bearing Thorns, Prickles, or Stiff Bristles*

1 Stems with thorns:

  2 Thorns in pairs at the nodes; leaves compound:

    3 Crushed foliage aromatic; leaflets showing translucent dots if held to the light . . . . prickly-ash (*Zanthoxylum americanum*) (p. 244)

    3 Crushed foliage not aromatic; leaflets without translucent dots . . . . black locust (*Robinia pseudoacacia*) (p. 242)

  2 Thorns not in pairs at the nodes:

    4 Leaves compound, with many small leaflets; thorns usually branched . . . . honeylocust (*Gleditsia triacanthos*) (p. 232)

    4 Leaves simple; thorns usually unbranched:

      5 Thorns often bearing leaves or buds:

        6 Leaves finely toothed, sometimes double toothed, with sharply pointed teeth . . . . American plum (*Prunus americana*) (p. 225)

        6 Leaves finely toothed with rounded or incurved teeth; upper surface of leaves waxy . . . . pear (*Pyrus communis*) (p. 211)

        6 Leaves coarsely or deeply toothed or lobed, or some leaves nearly entire . . . . crab apple (*Malus*) (p. 208)

      5 Thorns rarely if ever bearing leaves or buds:

        7 Leaves entire:

          8 Leaves 1–3 cm long, rounded . . . . Japanese barberry (*Berberis thunbergii*) (p. 170)

8 Leaves 7–15 cm long, tapering to long point . . . . .Osage-orange (*Maclura pomifera*) (p. 161)

7 Leaves toothed and often lobed . . . . . . hawthorn (*Crataegus*) (p. 204)

1 Stems with prickles:

9 Leaves simple, entire or slightly rough margined; prickles often stout and thornlike . . . . .greenbrier (*Smilax*) (p. 104)

9 Leaves simple, lobed; prickles mostly at the nodes . . . . .gooseberry (*Ribes*) (p. 189)

9 Leaves once compound, with three to nine leaflets:

10 Stipules attached to petiole for half their length or more, narrow but easy to see. . . . . .rose (*Rosa*) (p. 219)

10 Stipules attached to petiole only at base. . . . . blackberry, raspberry (*Rubus*) (p. 212)

9 Leaves mostly twice compound, with many leaflets; prickles borne on trunk, stems, and leaves . . . . devils-walkingstick (*Aralia spinosa*) (p. 296)

1 Stems with stiff bristles:

11 Leaves simple, entire . . . . .glaucous greenbrier (*Smilax glauca*) (p. 105)

11 Leaves simple, lobed, maplelike . . . . flowering raspberry (*Rubus odoratus*) (p. 213)

11 Leaves with three to seven leaflets . . . . wineberry (*Rubus phoenicolasius*) (p. 215)

red raspberry (*Rubus idaeus* var. *canadensis*) (p. 215)

11 Leaves with seven to nineteen leaflets . . . . bristly locust (*Robinia hispida*) (p. 240)

VII *Stems Green or Red, or Green Beneath and Red on Side Facing Light*

1 Stems green:

2 Leaves opposite:

3 Leaves compound, with usually three to five leaflets . . . . boxelder (*Acer negundo*) (p. 265)

3 Leaves simple, finely toothed; stems four-angled or four-sided . . . . euonymus (*Euonymus*) (p. 258)

3 Leaves simple, toothed and lobed . . . . maple (*Acer*) (p. 263)

3 Leaves simple, entire, the lateral veins curving forward and running nearly parallel to leaf margin . . . . flowering dogwood (*Cornus florida*) (p. 300)

2 Leaves alternate, aromatic if rubbed between thumb and forefinger:

4 Leaves entire or two- to three-lobed, with three principal veins from

near the base; stems usually bright green .... sassafras (*Sassafras albidum*) (p. 182)

    4 Leaves entire, with one principal vein from the base; size of leaves decreasing from tip to base of twig; stems dark green...spicebush (*Lindera benzoin*) (p. 182)

  2 Leaves alternate, small, 2–4 cm long, entire; stems very slender, speckled with warty dots; low shrub .... lowbush blueberry (*Vaccinium vacillans*) (p. 329)

1 Stems red, or partly red and partly green:

  5 Leaves opposite, simple, toothed and usually lobed .... red maple (*Acer rubrum*) (p. 269)

  5 Leaves alternate:

    6 Leaves with large glands at base of blade .... peach (*Prunus persica*) (p. 224)

    6 Leaves heart shaped, single toothed; buds usually some shade of red, mucilaginous if chewed .... basswood (*Tilia*) (p. 286)

    6 Leaves dotted beneath with resin droplets or patches; twigs pink to green .... huckleberry (*Gaylussacia*) (p. 321)

    6 Leaves large, averaging about 12 cm in length, more or less lance shaped, finely toothed, sour or bitter to taste .... sourwood (*Oxydendrum arboreum*) (p. 320)

    6 Leaves smaller, usually less than 8 cm long, minutely toothed; twigs speckled with whitish warty dots.................... highbush blueberry (*Vaccinium constablaei*) (p. 330)

VIII *Stems and Often the Leaves Aromatic*
(if rubbed or if the bark is broken)

1 Leaves simple:

  2 Leaves opposite, entire; twigs flattened at the nodes
....sweetshrub (*Calycanthus*) (p. 179)

  2 Leaves alternate:

    3 Leaves fernlike, dotted with yellow resin droplets; low shrub
.... sweet-fern (*Comptonia peregrina*) (p. 117)

    3 Stems marked with a faint ring (stipule scar) at each node:

      3a Leaves pointed at apex .... magnolia (*Magnolia*) (p. 174)

      3a Leaves broadly and very shallowly notched at apex ..... yellow-poplar (*Liriodendron tulipifera*) (p. 177)

    3 Stems with strong odor of bitter almonds when bruised...... cherry, peach (*Prunus*) (p. 223)

3 Stems with odor of wintergreen if bruised . . . . sweet birch, yellow birch
(*Betula*) (p. 129)
  3 Terminal bud large and prominent:
    3a Crushed buds with odor of balsam, often gummy; leaves toothed
    . . . . cottonwood (*Populus*) (p. 112)
    3a Crushed buds with spicy odor, not gummy; leaves entire or two- or
    three-lobed. . . . . sassafras (*Sassafras albidum*) (p. 182)
  3 Buds usually in clusters of two or more at a node; leaves obovate, entire
  . . . . spicebush (*Lindera benzoin*) (p. 182)
1 Leaves compound:
  4 Odor rank and unpleasant, likened to that of popcorn with rancid butter;
  twigs stout and blunt. . . . . ailanthus (*Ailanthus altissima*) (p. 245)
  4 Odor characteristically "nutty"; pith chambered; leaflets eleven to twenty-
  three . . . . walnut (*Juglans*) (p. 117)
  4 Odor fragrant and pleasant:
    4a Twigs stout; leaves with seven to nine leaflets. . . . . .mockernut hickory
    (*Carya tomentosa*) (p. 122)
    4a Twigs slender; leaves with nine to thirty-five leaflets, resembling those
    of black locust . . . . indigobush (*Amorpha*) (p. 236)

IX *Stems with Three or More Buds Clustered at Tips*

1 Leaves more or less evenly spaced along stems, or at least not noticeably
crowded; leaf blades mostly toothed or lobed . . . . oak (*Quercus*) (p. 142)
1 Leaves crowded toward the end of the season's growth; leaf blades entire or
essentially so (margin may be hairy):
  2 Lower surface of leaves with fringed glands along the midrib (visible
  under magnifying lens). . . . . minnie-bush (*Menziesia pilosa*) (p. 312)
  2 Lower surface of leaves without glands along midrib
  . . . . .azalea (*Rhododendron*) (p. 304)

Illustrated Descriptions of
Genera and Species

Pine Family (Pinaceae)

Trees containing resin; leaves evergreen (in our species), mostly needle-like, single or in bundles (fascicles); fruit a woody cone with seed-bearing scales; pine, spruce, hemlock, true fir.

Pinus (pine)

Trees with a single straight trunk and whorled, spreading branches.

Foliage leaves evergreen, needlelike, in our species in *fascicles* (bundles) of two, three, or five, each fascicle enclosed at base by a papery sheath, sooner or later deciduous.

Buds scaly, variable in size, shape, and color, with many overlapping scales, resinous or nonresinous.

Pollen-producing cones appearing briefly in spring.

Seed-bearing cones with woody scales, maturing in autumn of second year and commonly persisting after seeds are shed; seeds winged.

1 Leaves in fascicles of five, the sheath of each fascicle quickly deciduous
.... (1) *P. strobus*
1' Leaves in fascicles of two or three, the sheath of each fascicle persistent:
  2 Leaves mostly 6–14 cm long, in twos or threes:
    3 Leaves mostly in twos, sometimes in threes, slender, straight, soft and flexible, bluish green; cone scales armed with a weak prickle
.... (2) *P. echinata*
    3' Leaves in threes, stout, stiff, twisted, yellowish green; cone scales armed with a stout prickle .... (3) *P. rigida*
  2' Leaves mostly 2–6 cm long, in twos:
    4 Leaves 1 mm wide or less, firm or flexible, usually yellowish green; cone scales tipped with a small, slender prickle 1–3 mm long
.... (4) *P. virginiana*
    4' Leaves more than 1 mm wide (usually about 1.5 mm), stiff, dark green or bluish green; cone scales tipped with a stout upcurved prickle 3–6 mm long .... (5) *P. pungens*

FIGURE 4 | *1. Pinus strobus*
eastern white pine

A large tree. Bark on mature trunks deeply and closely furrowed into thick, roughly rectangular blocks having minute scales on surface.

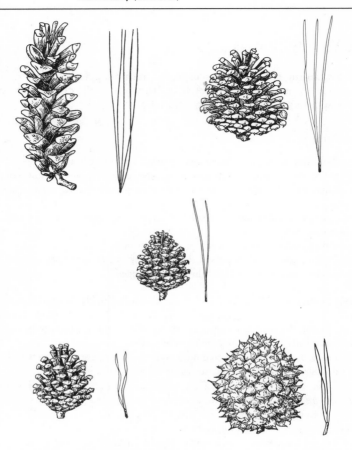

FIGURE 4   Cones and leaves of *Pinus:* Above: eastern white pine, *P. strobus* (left), pitch
pine, *P. rigida.* Center: shortleaf pine, *P. echinata.* Below: Virginia pine, *P.
virginiana* (left), Table Mountain pine, *P. pungens.*

Leaves 8–12 cm long, *in fascicles of five,* the base of each fascicle surrounded at
first with many scales, these soon falling off; leaves slender, straight, soft, *minutely toothed* (evident under magnifying lens), dark *bluish green* but whitened
along one side.
Twigs greenish or light greenish brown, glabrous or slightly hairy.
Buds with thin, reddish to orange brown scales.

Cones 10–15 cm long, cylindrical, long stalked; cone scales thin, rounded, flat, without prickles.

Dry or moist woods; common, throughout.

| | 2. *Pinus echinata* |
| FIGURE 4 | shortleaf pine |

A large tree. Mature bark purplish to reddish brown, divided into irregular, flat, scaly plates.

Leaves 7–13 cm long, *in fascicles mostly of two but also in threes on same tree,* slender, flexible, dark yellowish green.

Twigs slender, green tinged with purple, aging to reddish brown.

Buds with reddish brown scales, not or only slightly resinous.

Cones small, only 4–6 cm long, conical or narrowly ovoid, dull brown, with weak slender prickles on scales.

Dry rocky or sandy woods; fairly common, throughout.

| | 3. *Pinus rigida* |
| FIGURE 4 | pitch pine |

A medium-sized tree. Bark of trunk brown, broken into rough, irregular plates separated by deep, narrow furrows.

Leaves 5–14 cm long, mostly *in fascicles of three,* stiff, somewhat *twisted,* dark yellowish green, standing out at nearly right angles to twig.

Twigs stout, rough, grayish to purplish brown.

Buds conical, resinous, reddish brown.

Cones 4–8 cm long, with *stout rigid prickles* on scales.

Dry slopes and ridges at lower elevations; fairly common, throughout.

| | 4. *Pinus virginiana* |
| FIGURE 4 | Virginia pine |

Usually a small tree; lower branches persisting for a long time, even when dead. Bark of trunk in thin brownish orange scales.

Leaves mostly 2–6 cm long, *in fascicles of two,* stout, rather stiff, often twisted, yellowish green to grayish green, generally standing out at wide angle to twig.

Twigs slender, *tough and flexible,* covered with *glaucous bloom* when young, grayish brown when older.

Buds ovoid to oblong, pointed, very resinous.

Cones 3–6 cm long, oblong-conical, opening when mature but *remaining attached* to tree for several years; cone scales armed with sharp, *persistent* prickles.

Old or abandoned fields, dry sandy or rocky woods; common, throughout.

| | 5. *Pinus pungens* |
| | Table Mountain pine |

A small tree. Bark thick, dark reddish brown, broken into scaly plates.

Leaves 3–6 cm long, usually *in fascicles of two, stout,* rigid and *crowded,* sharp pointed, often twisted, dark green or bluish green.

Twigs stout, tough, persistent, light orange and smooth when young, brown and rough when older.

Cones *large,* 5–9 cm long, ovoid, heavy, often in dense clusters; scales *much thickened* at end, each armed with a *stout curved prickle;* cones opening late and persisting for many years.

Dry or rocky woods, high mountain ridges; fairly common, throughout.

| | *Picea rubens* |
| FIGURE 5 | red spruce |

A large tree. Bark of trunk thin, scaly, reddish brown.

Leaves spirally arranged, close set, bristling around twig in all directions, 10–15 mm long, needlelike, *four-angled* in cross section, sharp pointed, without petioles, borne on conspicuous peglike projections, dark green or dark yellowish green, shiny.

Twigs orange brown, more or less hairy, *roughened by many peglike stubs* that remain after leaves have fallen.

Buds ovoid, pointed, reddish brown, with many overlapping scales.

Cones hanging down, 2–6 cm long, woody, light reddish brown, with shiny, rigid scales; maturing in autumn of first year and usually falling after seeds are shed.

High mountain forests and balds, mainly above 4,000 feet; occasional in NC, common in GSMNP.

Tsuga (hemlock)

Medium-sized to large trees.

Leaves evergreen, arranged spirally, variable in length, those on upper side of twig usually shortest, needlelike, flattened, soft, bluntly pointed, with a *distinct* short petiole and *two whitish bands on lower side.*

Twigs very slender, *roughened by many small peglike projections* that bear leaves, but the projections smaller than those in *Picea.*

Cones woody, thin scaled, the scales persistent; borne at ends of lateral twigs of previous year; maturing in one year, remaining on tree for a long time.

1 Leaves less than 15 mm long, minutely toothed, more or less flattened in one plane on twigs; cones less than 2 cm long . . . . (1) *T. canadensis*

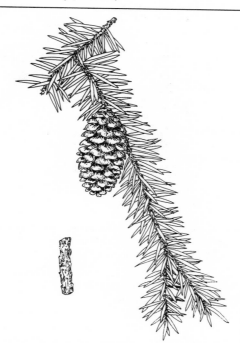

FIGURE 5   Branch and cone of red spruce, *Picea rubens*. Twig section showing peglike
projections that remain after leaves are shed.

1' Leaves mostly more than 15 mm long, entire, spreading in all directions on
   twigs; cones more than 2 cm long.........................(2) *T. caroliniana*

FIGURE 6 | 1. *Tsuga canadensis*
         | eastern hemlock

A medium-sized to large tree with gracefully spreading branches. Mature bark
   brown to purplish, deeply furrowed into scaly ridges.
Leaves 10–15 mm long, short petioled, flat, flexible, blunt, *minutely toothed* (under
   magnifying lens), shiny dark green above, *whitened in two lines beneath;* arranged spirally, but petioles bent so as to hold leaves in a more or less horizontal
   plane; usually *a row of upside-down leaves* along upper side of twig.
Cones 15–20 mm long, brownish to gray, few scaled, hanging down, long
   persistent.
Moist, cool ravines and valleys, rocky streambeds, north-facing bluffs; common,
   throughout.

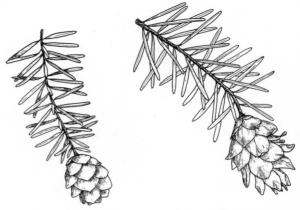

FIGURE 6   *Tsuga*. Left: branch cone of eastern hemlock, *T. canadensis,* showing characteristic upside-down leaves. Right: branch and cone of Carolina hemlock, *T. caroliniana.*

FIGURE 6	2. *Tsuga caroliniana* Carolina hemlock

A medium-sized tree. Bark thick, reddish brown, deeply furrowed into broad, flat scaly ridges.

Leaves 15–18 mm long, flat, flexible, *entire,* slightly notched at tip, shiny dark green above, whitish beneath, *set at various angles along twigs* and more thickly distributed, giving branches a rougher appearance than in *T. canadensis.*

Twigs very slender, orange brown, slightly hairy, *rough with peglike stubs.*

Cones 25–35 mm long, narrowly elliptical, yellow to light brown, persisting for only a few months after seeds are shed.

South-facing slopes, dry rocky ledges and cliffs; occasional, throughout.

FIGURE 7	*Abies fraseri* Fraser fir

A small to medium-sized tree. Bark gray or brown, rather smooth, with *many resin blisters;* older bark with thin, papery scales.

Leaves alternate, spirally arranged but twisted at base to appear two-ranked, evergreen, linear or needlelike, 1–2 cm long, flattened, blunt or notched at tip, shiny green above, *whitish beneath;* each leaf with *broad circular base.*

Twigs stout, hairy, bearing *circular* leaf scars, the scars *flat* against twig (not raised) after leaves are shed or removed.

FIGURE 7   Foliage and cone of Fraser fir, *Abies fraseri*.

Buds nearly globose, resinous, with orange green scales.

Cones *erect,* 4–6 cm long, dark purple, their scales deciduous in first autumn, leaving axis persistent on branch.

Spruce-fir forests and balds at high elevations, mainly above 5,000 feet; NC, GSMNP.

Cypress Family (Cupressaceae)

Evergreen trees or shrubs; leaves usually opposite, occasionally whorled, scalelike and overlapping or needlelike and spreading, sometimes both forms on same plant; fruit a leathery, woody, or fleshy cone; white-cedar, juniper.

FIGURE 8	*Thuja occidentalis* northern white-cedar

A medium-sized tree with reddish brown, shreddy bark on old trunks.

Leafy twigs yellowish green, *strongly flattened* and arranged in fanlike sprays. Leaves very small, *scalelike, tightly appressed, in alternate pairs,* those on upper and lower surfaces flat, the lateral pairs folded.

Cones *small,* 1–2 cm long when mature, elliptical, with light brown scales; maturing in first year.

Moist or wet woods; Alleghany, Ashe, Burke, Haywood Cos., NC; GSMNP.

FIGURE 8	*Juniperus virginiana* eastern redcedar

An erect tree or shrub. Bark of trunk light brown to reddish brown, *thin, fibrous, shreddy,* peeling in long narrow strips.

Leaves evergreen, yellowish green to bluish green, of two kinds: (*a*) typically scalelike, 2 mm long or less, opposite, *closely appressed and overlapping,* in four rows lengthwise on twigs, making twigs four-sided in cross section; and (*b*) on very young trees and often with typical form on older trees, more or less *needlelike,* 6–12 mm long, *sharp pointed,* spreading, opposite or in whorls of three, not overlapping.

Leafy twigs slender, four-angled, not having a flattened, fanlike appearance as in *Thuja.*

Buds minute, naked, covered by leaves.

Fruits *fleshy, berrylike,* 10–13 mm across, dark blue under light blue waxy bloom.

Dry woods, abandoned fields, fencerows; infrequent, essentially throughout.

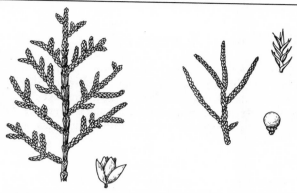

FIGURE 8   Left: branch and cone of northern white-cedar, *Thuja occidentalis*. Right: eastern redcedar, *Juniperus virginiana*. Twigs showing two forms of leaves and "berry."

Lily Family (Liliaceae)

A large family of mostly herbaceous plants, only a few members woody; stems unbranched or sparingly branched, with numerous scattered vascular bundles (lacking a cambium layer) and hence no annual rings; leaves generally alternate, simple; flowers typically showy; fruit a berry or capsule. In our area the following genus includes woody vines.

Smilax (greenbrier, catbrier)

Woody vines *climbing by means of paired tendrils* (modified stipules) attached near base of leaf petioles. Some species herbaceous, but those not considered here.

Leaves alternate, deciduous or evergreen, simple, entire or slightly rough margined, *the principal veins curving and nearly parallel* from base to apex of blade.

Stems circular or four-angled in cross section, usually *green; central pith lacking,* the vascular bundles scattered throughout stem (evident if stem is carefully sectioned crosswise); usually armed with *few to many prickles,* these well distributed or confined mainly to base of plant.

Buds slender, three-sided, pointed, with only one scale, concealed by partly clasping base of leaf petiole.

Leaf scars usually not visible, the leaves breaking or tearing off above base, a part of old petiole remaining, covering the true leaf scar.

Fruit a small black or bluish black berry, often persistent into winter.

1 Leaves strongly bluish silvery or whitened beneath; leaf margin entire
   .... (1) *S. glauca*
1' Leaves green beneath, not bluish silvery or whitened, or only slightly so:
   2 Leaves variable in shape, at least some narrowed above base and hence
      fiddle shaped or three-lobed; leaf margin thickened, often spiny or bristly
      .... (2) *S. bona-nox*
   2' Leaves broadly ovate to lanceolate, not or rarely narrowed above base;
      leaf margin not thickened, entire or minutely spiny toothed:
      3 Prickles slender, needlelike or bristlelike, greenish at first, later black,
         and of different lengths; stems circular or only slightly angled
         .... (3) *S. hispida*
      3' Prickles stout, definitely flattened, mostly green with brownish or black-
         ish tip; stems angled, the prickles confined to the angles
         .... (4) *S. rotundifolia*

| FIGURE 9 | 1. *Smilax glauca*
glaucous greenbrier, catbrier |

A vine, often forming dense tangles.

Leaves deciduous, or many leaves persisting through winter, nonleathery but firm, 5–9 cm long, narrowly to broadly ovate or occasionally lanceolate, blunt or short pointed at apex, straight across or rounded at base, glossy green above, *distinctly whitish or bluish beneath;* leaves usually reddish on upper side in winter.

Stems slender, tough and wiry, *rounded,* green or brown, glaucous, armed with a few short curved prickles, or prickles more numerous near base of plant; prickles mostly green with a dark tip.

Berries globose, 5–9 mm across, bluish black with bloom, often persistent.

Open woods, roadsides, old fields, thickets; common, throughout.

| FIGURE 9 | 2. *Smilax bona-nox*
sawbrier, catbrier |

A vine, climbing or sprawling over low bushes.

Leaves deciduous or partly evergreen, firm and somewhat leathery, 4–10 cm long, widest at base, *often constricted near middle,* and varying from heart shaped to somewhat *triangular or fiddle shaped,* green and shining on both surfaces; margin thickened and frequently spiny or bristly.

Stems sharply *four-angled,* zigzag, usually armed with prickles, green, glabrous. Prickles confined to angles of stem, stiff, flat, broad based, entirely green or often with a dark tip.

Berries 6–8 mm across, black, usually one-seeded.

Dry woods, thickets, old fields, roadsides; Buncombe, Madison, Transylvania Cos., NC; Oconee Co., SC; GA.

| FIGURE 9 | 3. *Smilax hispida*
bristly greenbrier |

A high-climbing vine.

Leaves deciduous or often retained through winter, thin, mostly ovate but varying from lanceolate to nearly circular, 6–12 cm long, slightly rough margined or minutely toothed (under magnifying lens), *dark green* and lustrous on both surfaces.

Stems slender, *rounded,* green, glabrous. Prickles *straight, sharp but slender, bristly,* green to straw colored at first, later blackish; prickles few or none on young stems, numerous toward base of plant.

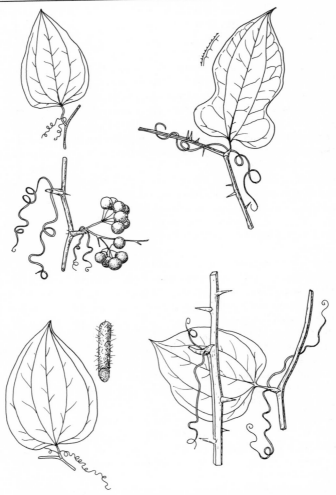

FIGURE 9   *Smilax.* Upper left: glaucous greenbrier, *S. glauca,* leaf, stem, and fruit.
Upper right: sawbrier, *S. bona-nox,* leaf and stem. Lower left: bristly
greenbrier, *S. hispida,* leaf and stem. Lower right: common greenbrier, *S.
rotundifolia,* leaf, stem, and section of stem showing prickles.

Berries globose, 6–8 mm across, black without bloom.
Moist woods and thickets; occasional, throughout.

| FIGURE 9 | 4. *Smilax rotundifolia*
common greenbrier |

A vine, often high climbing or forming dense tangles.

Leaves deciduous or partly evergreen, firm, usually ovate but ranging from nearly
   circular to broadly lanceolate, mostly 5–10 cm long, entire or minutely rough
   margined (under magnifying lens), *green* and shiny on both sides.

Stems green, prominently to obscurely *four-angled,* glabrous, not glaucous. Prickles
   stout, strong, flattened, green at base, dark brownish to blackish at tip, confined
   to angles of stem, lacking at the nodes.

Berries 5–8 mm across, black or bluish black, glaucous.

Moist woods, thickets, roadsides; common, throughout.

Willow Family (Salicaceae)

Trees or shrubs; leaves alternate, simple, deciduous; bark bitter; flowers in catkins, early in spring, staminate and pistillate flowers on different plants; fruit a capsule containing many small tufted seeds; willow, poplar, cottonwood.

Salix (willow)

Shrubs or trees.

Leaves alternate, deciduous, simple, short petioled, *relatively narrow* in our species, usually toothed.

Twigs thin and limber but commonly *brittle at base,* easily broken off at point of attachment to older stem, often highly colored, with *very bitter bark.*

Buds covered by *only one scale,* usually flattened against twig. Terminal bud absent, twig tips dying back to uppermost lateral bud.

Leaf scars raised, narrow, constricted between three bundle scars. Stipules persistent or early deciduous; stipule scars present or absent.

Flowers and fruits (small capsules) in catkins; seeds bearing tufts of cottony or silky hairs; appearing early in growing season.

1 Lower surface of leaves green and lustrous, not or only slightly paler than upper; twigs glabrous or nearly so .... (1) S. *nigra*
1' Lower surface of leaves much paler than upper, decidedly glaucous or whitish-hairy:
  2 Leaves mostly entire or wavy margined; teeth, if present, rounded and uneven or far apart; margin usually slightly rolled under.......... (2) S. *humilis*
  2' Leaves regularly toothed; margin flat, not rolled under:
    3 Petioles usually with glands at base of blade; young twigs greenish yellow, hairy to nearly glabrous .... (3) S. *alba*
    3' Petioles without glands:
      4 Twigs glabrous or only thinly hairy, often lustrous; branches and twigs conspicuously long and drooping.............. (4) S. *babylonica*
      4' Twigs silky, short-hairy, or woolly; branches and twigs spreading or erect, never drooping:
        5 Leaves with their length more than four times the width; upper surface glabrous or practically so, lower surface more or less silvery-silky .... (5) S. *sericea*
        5' Leaves with their length two to three times the width; both surfaces grayish-hairy................................... (6) S. *cinerea*

FIGURE 10	1. *Salix nigra* black willow

A shrub or tree, often with more than one trunk. Bark of old trunks thick, very
  rough, often shaggy, dark brown to nearly black.
Leaves *narrowly lanceolate,* mostly six or more times longer than wide, 6–15 cm
  long, often curved like a scythe, *very finely toothed,* extending to a long slender
  tip at apex, rounded or broadly wedge shaped at base, *green and somewhat
  lustrous* on both sides, glabrous beneath or sometimes hairy along lower veins.
Twigs slender, flexible but *easily broken at base,* greenish to yellowish brown or
  dark brown, glabrous or slightly hairy toward tip.
Buds *small,* usually 3 mm long or less, reddish brown, glabrous.
Stipules winglike, often persisting on vigorous stems until autumn.
Streambanks, low woods; common, essentially throughout.

FIGURE 10	2. *Salix humilis* upland willow, prairie willow

A small shrub, usually with many stems from the base.
Leaves tending to be crowded toward tip of stem, oblanceolate, obovate, or elliptical, 2–11 cm long, *margin entire or wavy* or with a few blunt teeth, but not
  definitely and regularly toothed, often slightly curled under; short pointed or
  blunt at apex, wedge shaped to rounded at base, dull green and glabrous or hairy
  above; *lower surface* much paler than upper, *strongly glaucous,* more or less
  hairy but not white-silky.
Twigs moderately slender, purplish to yellowish brown or brown, usually *covered
  with ashy gray hairs.*
Buds *closely spaced,* 2–6 mm long, elliptical in outline, flattened, with broad blunt
  tips, light tan to reddish, hairy.
Stipules present or absent.
Balds, dry woods, openings, roadsides; infrequent, NC, GA, GSMNP.

FIGURE 10	3. *Salix alba* white willow

A medium-sized to large tree.
Leaves lanceolate, 2–10 cm long, finely toothed, long pointed at apex, wedge
  shaped at base, silky on both sides when young, becoming green and nearly glabrous above, remaining *silky-hairy and whitish beneath.* Petioles, at least some,
  bearing *small glands* at juncture with blade.
Twigs moderately slender, *yellow to greenish* when young, aging to reddish brown,

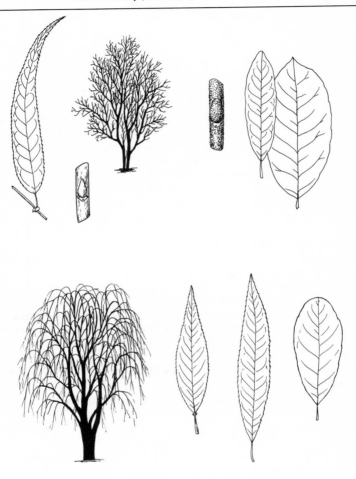

FIGURE 10   *Salix.* Above left: black willow, *S. nigra,* leaf, bud, and silhouette. Above right: upland willow, *S. humilis,* lateral bud and two forms of leaves. Below, left to right: silhouette of mature weeping willow, *S. babylonica,* and leaves of white willow, *S. alba,* silky willow, *S. sericea,* and gray willow, *S. cinerea.*

silky to nearly glabrous; flexible, not easily broken off at base, and often slightly drooping.

Buds small, usually less than 5 mm long, same color as twigs, commonly with some pale hairs.

Stipules absent or very small and early deciduous.

Native of Europe, occasionally escaped from cultivation.

| FIGURE 10 | 4. *Salix babylonica*
weeping willow |

A tree with gracefully *arching and drooping* branches.

Leaves narrowly lanceolate, 6–14 cm long, finely toothed, long pointed at apex, wedge shaped at base, dark green above, *whitish or gray and often silky beneath;* petioles without glands.

Twigs slender, olive green when young, yellowish or brownish at maturity, glabrous, *long and "weeping,"* brittle at base, easily broken off.

Buds 3–5 mm long, oblong, rounded at tip, yellowish to brown, glabrous or essentially so.

Native of China; rare escape, propagated from broken twigs.

| FIGURE 10 | 5. *Salix sericea*
silky willow |

A shrub with clustered stems from base.

Leaves mostly lanceolate, 2–14 cm long, finely and regularly toothed, long pointed at apex, narrowly to broadly wedge shaped at base, dark green and glabrous or nearly so on upper side; lower surface glaucous, *silky and glistening* when young, usually remaining silky, at least along veins.

Twigs slender, brittle at base, brown to purplish, usually short-hairy.

Buds 2–6 mm long, oblong with rounded or pointed tips, greenish, red, or brown, finely hairy.

Stipules slender, usually deciduous.

Low woods, streambanks; common, throughout.

| FIGURE 10 | 6. *Salix cinerea*
gray willow |

A shrub or occasionally treelike.

Leaves obovate or obovate-lanceolate, 2–9 cm long, very finely or obscurely toothed to almost entire, more or less pointed at apex, wedge shaped to rounded at base, dull green and somewhat hairy above, *densely grayish-hairy beneath.*

Twigs dark to *nearly black,* covered with pale matted wool.

Buds 3–6 mm long, permanently woolly.

Stipulates small or absent.

Native of Eurasia, rarely escaped from cultivation; Buncombe, Henderson Cos., NC.

Populus (cottonwood, poplar)

Fast-growing trees with *brittle,* easily detached twigs and *sharply bitter bark.*

Leaves alternate, deciduous, simple, *broad, long petioled,* toothed and occasionally lobed.

Twigs stout to slender, circular or somewhat angled in cross section. Short *spur twigs* usually present, roughened by raised, crowded leaf scars. Pith *five-angled* in cross section. Scars left by fallen bud scales clearly visible at base of annual growth.

Terminal bud present, *large,* with many scales exposed. Lateral buds with *lowest scale centered directly above leaf scar.*

Leaf scars raised, somewhat three-lobed, crescent shaped, or triangular. Bundle scars three, each sometimes divided. Stipule scars present, distinct, narrow.

Fruit a capsule containing a number of tufted seeds, borne in long catkins, only on pistillate trees; maturing generally in spring.

Summer Key

1 Leaves white-woolly on lower surface, sometimes lobed . . . . (1) *P. alba*
1′ Leaves not white-woolly, not lobed:
  2 Leaves with large teeth, fewer than fifteen on each side of midrib
  . . . . (2) *P. grandidentata*
  2′ Leaves with relatively small teeth, more than fifteen on each side of midrib:
    3 Petioles hairy; first-year twigs reddish brown . . . . (3) *P. gileadensis*
    3′ Petioles glabrous; first-year twigs yellow, yellowish brown, or orange:
      4 Leaves 5 – 10 cm long, usually wider than long, without glands at
      base of blade; branches ascending or erect, forming a very narrow
      crown . . . . (4) *P. nigra* var. *italica*
      4′ Leaves 8 – 18 cm long, usually longer than wide, with glands at base
      of blade; branches widely spreading, forming a broad crown
      . . . . (5) *P. deltoides*

Winter Key

1 Buds slightly or not at all resinous; terminal bud mostly less than 10 mm
long (poplars and aspens):
  2 End of twigs coated with a feltlike layer of whitish hairs, easily rubbed
  away . . . . (1) *P. alba*
  2′ End of twigs hairy or without hairs, but not coated with whitish hairs:
    3 Buds glabrous; twigs yellow, yellowish brown, or greenish brown
    . . . . (4) *P. nigra* var. *italica*
    3′ Buds somewhat hairy; twigs brownish gray . . . . (2) *P. grandidentata*
1′ Buds heavily resinous and gummy; terminal bud more than 10 mm long
(cottonwoods):
  4 Twigs yellowish brown or brownish yellow, glabrous . . . . (5) *P. deltoides*
  4′ Twigs reddish brown, dark brown, or olive brown, usually hairy
  . . . . (3) *P. gileadensis*

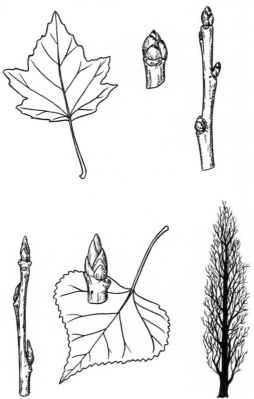

FIGURE 11   *Populus.* Above: white poplar, *P. alba,* leaf, terminal bud, and twig. Below: Lombardy poplar, *P. nigra* var. *italica,* twig, terminal bud, leaf, and silhouette of mature tree.

FIGURE 11	1. *Populus alba* white poplar

Bark smooth, light gray to greenish white, becoming dark and very rough on old trunks.

Leaves somewhat triangular in outline, 3–8 cm long, thick and firm, with a few coarse teeth or occasionally lobed, blunt at apex, straight across or slightly heart shaped at base, dark green above, *heavily felted beneath with a dense layer of whitish hairs.*

Twigs coated, at least near tip, with *whitish wool,* easily rubbed off, greenish or brown beneath the woolly coating.

Terminal but ovoid or conical, 5–7 mm long, more or less hairy, not resinous or
   only slightly so.
Native of Eurasia, sparingly escaped from cultivation.

| FIGURE 12 | 2. *Populus grandidentata*
bigtooth aspen |

Bark of trunk thin, smooth, olive green to grayish green, becoming thick, rough,
   and dark brown on old trunks.
Leaves broadly ovate to circular-ovate, 5–11 cm long and nearly as wide, *coarsely*
   and irregularly *toothed,* short pointed at apex, rounded at base, dull green
   above, paler beneath; *petioles flattened.*
Twigs moderately stout, dull brownish, hairy when young.
Terminal bud ovoid, 3–7 mm long, brown, gray-hairy to nearly glabrous, slightly
   resinous. Lateral buds much narrower than terminal buds, about the same
   length, more or less divergent.
Dry woods; Ashe, Haywood, Transylvania, Watauga Cos., NC; GSMNP.

| FIGURE 12 | 3. *Populus gileadensis*
balm-of-Gilead poplar |

Leaves heart shaped to broadly ovate, 5–17 cm long, finely toothed, *the teeth ciliate;* heart shaped to straight across at base, short pointed or long pointed at
   apex, dark green and glabrous above, much paler, glabrous or sparingly hairy
   and lustrous beneath; petioles *hairy,* rounded or at least not strongly flattened,
   often grooved on upper side.
Twigs dark brown, reddish brown, or olive brown, hairy at first, becoming glabrous
   and lustrous.
Terminal bud *very large,* mostly about 2 cm long, tapered to a sharp point, brown,
   glabrous, *resinous and sticky,* fragrant. Lateral buds narrow, appressed or tips
   divergent, with mostly three scales showing.
A hybrid of unknown origin, rarely spreading from cultivation by sprouts; NC, GA,
   GSMNP.

| FIGURE 11 | 4. *Populus nigra* var. *italica*
Lombardy poplar |

A tall tree, *very narrow* for its height, with *branches upright and crowded.*
Leaves triangular, 5–10 cm long, *usually wider than long,* as if stretched sideways,
   finely and regularly toothed, the teeth blunt and *without cilia,* abruptly long
   pointed at apex, broadly wedge shaped to straight across at base, bright green

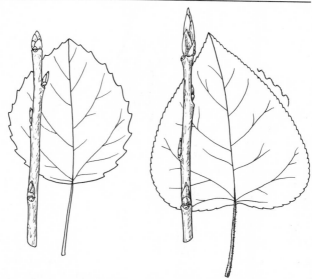

FIGURE 12   *Populus*. Left: bigtooth aspen, *P. grandidentata,* leaf and twig. Right: balm-of-Gilead poplar, *P. gileadensis,* leaf and twig.

above, paler beneath; petioles *flattened,* glabrous, without glands where blade and petiole meet.

Twigs moderately slender, yellowish brown or olive brown, glabrous.

Terminal bud 8–10 mm long, sharply pointed, angled in cross section, glabrous, *slightly resinous but not gummy.* Lateral buds usually less than 8 mm long, appressed, somewhat curved at tip, brown, glabrous.

A horticultural form of *P. nigra,* native of Eurasia; sparingly spread by root sprouts from plantings along roadsides and around old homesites.

FIGURE 13	**5. *Populus deltoides*** eastern cottonwood

Bark on old trunks grayish, very thick, deeply furrowed.

Leaves triangular or occasionally heart shaped, 8–18 cm long, glabrous, coarsely toothed, the teeth with hard, incurved tips; petioles glabrous, *flattened, with two glands at point where blade and petiole meet.*

Twigs stout, brittle at base, easily detached, yellowish brown or brownish yellow, often tinged with green, glabrous, lustrous, rounded or *three-ridged below nodes;* bark very bitter; lenticels large, pale, elongated longitudinally.

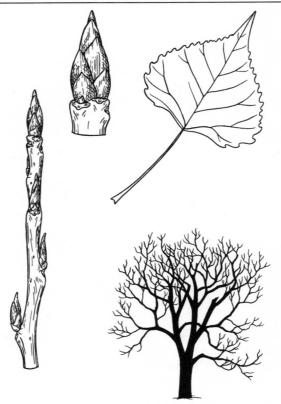

FIGURE 13   Eastern cottonwood, *Populus deltoides*. Left to right: twig, enlarged
terminal bud, leaf, and silhouette of mature tree.

Terminal bud *large,* 10–20 mm long, angled, broadest near middle, slightly narrowed at base, slender pointed, yellow to yellowish brown, glabrous, shiny, *resinous, strong scented,* with varnishlike odor when crushed; visible bud scales six
or seven. Lateral buds 10–15 mm long, divergent, with three or four scales exposed, the lowest scale directly above leaf scar.

Streambanks, bottomlands; doubtfully native in our area, probably planted; Watauga Co., NC; GA; GSMNP.

Sweet Gale Family (Myricaceae)

Shrubs or small trees; leaves alternate, deciduous or half-evergreen, simple, bearing many resin dots, aromatic; flowers in catkins; fruit a small drupelike nut. Represented in our area by one species.

FIGURE 14	*Comptonia peregrina* sweet-fern

A low shrub, often in colonies.

Leaves alternate, deciduous, simple, *fernlike,* linear-lanceolate to linear-oblong, 6–12 cm long, *deeply and evenly divided into fifteen to twenty-five lobes,* resin dotted on both surfaces, *aromatic* when crushed, dark green above, paler beneath, more or less hairy.

Twigs slender, brown to purplish brown, covered with white hairs, bearing minute *yellow resin droplets when young* (under magnifying lens), fragrant if cut or broken.

Terminal bud absent. Lateral buds small, globe shaped, with four or more scales visible.

Staminate catkins exposed in winter, erect, clustered near ends of twigs, brown, with many resin-dotted scales.

Leaf scars raised, triangular to three-lobed. Bundle scars three, distinct. Stipule scars present, small.

Fruit a smooth, ovoid, bony nutlet, 3–4 mm long, several together in a burlike head.

Open, dry woods and roadsides; fairly common, NC, SC, GA.

Walnut Family (Juglandaceae)

Deciduous trees with stout twigs; large terminal bud present; leaves alternate, deciduous, pinnately compound; staminate flowers borne in catkins, pistillate in erect spikes (reduced catkins); fruit a drupelike nut enclosed in a husk; walnut, hickory.

Juglans (walnut)

Trees with pinnately compound leaves and *chambered pith.*

Leaves alternate, deciduous, with eleven to twenty-three finely toothed leaflets, aromatic when crushed.

FIGURE 14   Sweet-fern, *Comptonia peregrina*. Twig with leaves, catkins, and leaf buds.

Twigs *stout,* circular in cross section. Pith chambered after first season, consist-ing of empty chambers alternating with thin partitions, clearly visible if twig is cut lengthwise.

Terminal bud present, large, appearing naked but actually covered by two to four thick, hairy scales, the outer scales as long as the bud. Lateral buds smaller, often one above another (superposed).

Leaf scars large, prominent, more or less three-lobed. Bundle scars numerous, in *three clearly defined, U-shaped groups.* Stipules and their scars absent.

Fruit a *rough nut* enclosed in thick, fleshy, *nonsplitting husk,* solitary or in clus-ters of three to five; kernel sweet, edible.

1 Terminal bud conical-oblong, 12–18 mm long; twigs with a pad of feltlike
   hairs in upper angle between twig and leaf petiole (or in the absence of
   leaves, hairy pad present along upper margin of leaf scar)......... (2) *J. cinerea*
1' Terminal bud broadly ovoid, 7–10 mm long; twigs without a pad of feltlike
   hairs in upper angle between twig and leaf petiole (and hairy pad similarly
   lacking along upper margin of leaf scar) .... (1) *J. nigra*

	1. *Juglans nigra*
FIGURE 15	black walnut

A large tree with heavy erect or somewhat spreading branches. Mature bark *dark brown to grayish black, very rough* with scaly ridges.
Leaflets fifteen to twenty-three (or fourteen to twenty-two; terminal leaflet often

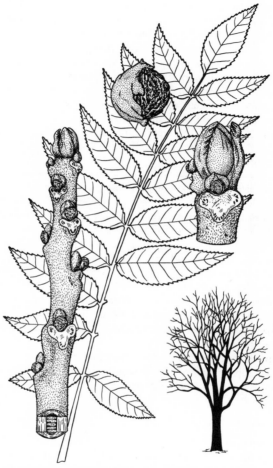

FIGURE 15   Black walnut, *Juglans nigra*. Twig showing terminal bud, superposed lateral buds, leaf scars, and chambered pith; compound leaf; fruit with husk partially removed; enlarged terminal bud, topmost lateral bud, leaf scar, and U-shaped bundle scars; silhouette of mature tree.

missing), lanceolate to narrowly ovate, 3–15 cm long, finely toothed, short pointed or long pointed at apex, rounded, heart shaped, or unequal at base, dark yellowish green and glabrous above, pale green and hairy beneath; crushed leaves somewhat fragrant; petioles hairy to nearly glabrous.

Twigs orange brown or greenish brown to brown, hairy to nearly glabrous; pith *light brown,* chambered, the partitions thinner than in *J. cinerea.*
Buds woolly with grayish to brownish hairs. Terminal bud broadly ovoid, usually less than 10 mm long, covered with two pairs of scales but appearing naked. Lateral buds much smaller, often superposed, the upper bud larger than lower.
Leaf scars with *upper margin prominently notched, without a hairy pad.*
Fruits usually solitary or in pairs, globose or nearly so, mostly 4–5 cm across, the husk smooth or slightly rough, *not sticky-hairy.*
Rich woods; occasional, throughout.

	2. *Juglans cinerea*
FIGURE 16	butternut

A medium-sized tree. Bark of mature trunks divided into *broad, flat, shiny ridges;* bark of branches and younger trunks *light gray,* almost silvery.
Leaflets eleven to seventeen, terminal leaflets usually present, oblong-lanceolate, 3–17 cm long, finely toothed, long pointed at apex, unequal or rounded at base, yellowish green above, paler and softly hairy beneath; petioles resinous-hairy; foliage more clammy and fragrant than that of *J. nigra.*
Twigs olive green to brown, glandular-hairy at first, later glabrous. Pith *dark brown,* chambered.
Buds covered with pale, grayish brown hairs. Terminal bud longer than broad, usually more than 12 mm long, blunt, obliquely flattened, few scaled, often appearing naked. Lateral buds much smaller, frequently superposed, the upper bud the larger of the two.
Leaf scars with *transverse hairy pad along upper margin.*
Fruits in groups of three to five, *oblong,* 4–7 cm long, *husk sticky-hairy.*
Rich woods; occasional, throughout.

Carya (hickory)

Deciduous trees.
Leaves alternate, pinnately compound; leaflets five to eleven (in our species), usually fewer than in the walnuts, toothed and mostly long pointed, more or less fragrant when crushed.
Twigs stout to slender, tough and flexible, not easily broken; pith solid, *not chambered,* angled in cross section.
Buds sometimes superposed, rather *large,* the terminal bud largest, with overlapping scales (valvate in *C. cordiformis*).
Leaf scars alternate, *large, conspicuous,* shield shaped or three-lobed. Bundle scars *numerous,* scattered or in three or more groups. Stipule scars none.
Fruit a hard-shelled *nut,* enclosed in an *outer husk* partly or completely splitting lengthwise into four parts when ripe.

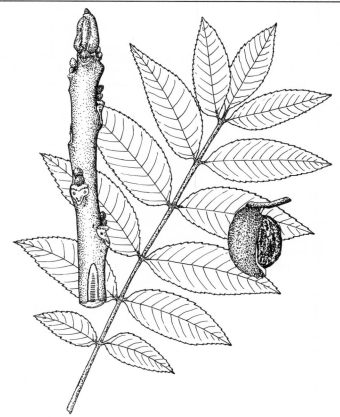

FIGURE 16   Butternut, *Juglans cinerea.* Twig showing terminal bud, superposed lateral
buds, leaf scars, and chambered pith; compound leaf; fruit with portion of
husk removed.

1 Buds yellow; terminal buds narrow, rounded or angled on sides, appearing
  naked but actually with valvate scales .... (1) *C. cordiformis*
1′ Buds not yellow; terminal buds not as above, but with overlapping scales:
  2 Terminal buds, all or most, more than 12 mm in length; twigs stout:
    3 Leaflets seven to nine, rarely five; young twigs and leaf petioles woolly
      with tufts of curly hairs; terminal buds with outer scales usually shed in
      autumn, exposing smooth silky scales beneath......... (2) *C. tomentosa*
    3′ Leaflets five, rarely seven; young twigs and leaf petioles less hairy or
      glabrous; terminal buds with outer scales usually persistent, often ex-
      tended into long, rigid points .... (3) *C. ovata*

2' Terminal buds less than 12 mm in length, mostly less than 10 mm, the outer scales soon deciduous, showing lighter-colored inner scales; twigs rather slender:

4 Terminal buds usually less than 7 mm long; outer scales of terminal bud abundantly dotted with silvery or golden scaly particles, visible under magnifying lens; lower surface of leaflets similarly dotted; petioles covered with tufts of hairs . . . . (4) *C. pallida*

4' Terminal buds 8–11 mm long; outer scales of terminal bud and lower surface of leaflets without silvery or golden scaly particles, or these only few or scattered; petioles without tufts of hairs

. . . . (5) *C. glabra*

FIGURE 17 | **1. *Carya cordiformis***
bitternut hickory

A tall, slender tree. Bark of trunk dark gray, smooth when young, later divided into a network of flat, narrow ridges, remaining tight, never shaggy.

Leaflets usually nine (seven to eleven), lanceolate or ovate-lanceolate to oblong-lanceolate, mostly 7–15 cm long, coarsely toothed, long pointed at apex, wedge shaped to rounded at base, yellowish green and usually glabrous above, pale green, hairy, and often gland dotted beneath.

Twigs slender to moderately stout, three-ridged below nodes, greenish brown to light brown, scurfy near tip but otherwise glabrous.

Terminal bud 6–18 mm long, *narrow and compressed,* obliquely blunt pointed, *yellow,* with rough or pitted surface. Bud scales *in pairs, valvate,* the outer pair often leaflike, about as long as bud and nearly covering it. Lateral buds smaller, commonly two together, one above the other (superposed), *often stalked.*

Nuts 25–35 mm long, including *thin, four-winged husk,* nearly globe shaped, thin shelled, kernel bitter.

Rich, low or upland woods; common, throughout.

FIGURE 18 | **2. *Carya tomentosa***
mockernut hickory

A fairly large tree with gray, furrowed, *close-fitting bark,* never shaggy; foliage fragrant when bruised.

Leaflets seven to nine, oblong-lanceolate to narrowly obovate, 10–20 cm long, the uppermost leaflet largest, finely to coarsely toothed, yellowish green and somewhat lustrous above, the lower surface pale green tinted with orange brown; *petioles and lower side of leaflets covered with tufts of woolly hairs.*

Twigs stout, reddish brown to brownish gray, with clustered hairs, giving off a strong "nutty" odor if cut or bruised.

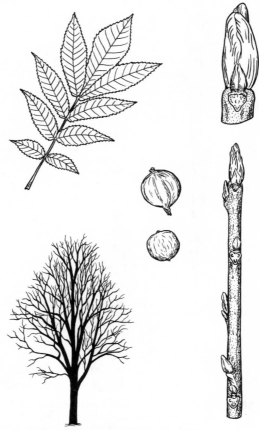

FIGURE 17   Bitternut hickory, *Carya cordiformis*. Compound leaf, fruit enclosed in husk and with husk removed; enlarged terminal bud; twig showing leaf scars, superposed buds, some buds stalked; silhouette of mature tree.

Terminal bud *large,* 1–2 cm long, broadly ovoid, bluntly pointed, light brown to dark brown, hairy, the *outer scales soon deciduous, exposing smooth silky scales beneath*. Lateral buds similar but smaller, somewhat divergent.

Fruits 4–5 cm long, including *thick husk,* nearly globe shaped, little flattened, thick shelled, kernel sweet and edible.

Upland woods; common, throughout.

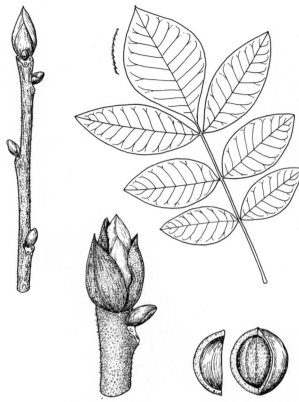

FIGURE 18   Mockernut hickory, *Carya tomentosa*. Twig with smooth inner scales of terminal bud showing, the outer scales having fallen off; enlarged terminal bud with outer scales persisting; compound leaf; fruit with a portion of husk removed.

FIGURE 19	3. *Carya ovata* shagbark hickory

A large tree. Bark at first smooth, soon becoming shaggy, *breaking loose in long strips that curl outward at ends.*

Leaflets typically five (five to seven), ovate or obovate, 7–20 cm long, the upper three leaflets largest, toothed, short pointed or long pointed at apex, wedge shaped at base, dark yellowish green above, pale yellowish green beneath, nearly or entirely without hairs except for persistent *tufts of hairs on teeth.*

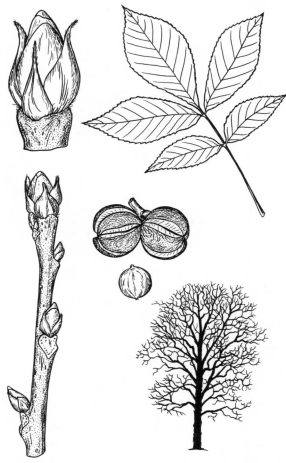

FIGURE 19 Shagbark hickory, *Carya ovata*. Enlarged terminal bud showing outer scales extended into long points; compound leaf; twig with terminal bud, lateral buds, and leaf scars; fruit enclosed in husks (splitting), and nut; silhouette of mature tree.

Twigs stout, reddish brown to grayish brown or blackish brown, hairy or glabrous; lenticels orange, conspicuous, elongated longitudinally.

Terminal bud *large,* often more than 15 mm long, ovoid, bluntly pointed. Bud scales three or four visible, the outer scales loose fitting but persistent, dark brown to blackish brown or dark gray, short-hairy, keeled, the *tips usually prolonged into*

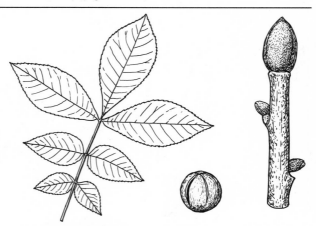

FIGURE 20   Sand hickory, *Carya pallida*. Compound leaf, nut enclosed in husk, and enlarged twig showing inner scales of terminal bud after outer scales have been shed.

slender, rigid points; inner scales brownish gray, short-hairy. Lateral buds similar but much smaller, divergent.

Fruits 4–6 cm in diameter, including *thick husk,* slightly flattened, four-angled or four-ribbed, kernel sweet and delicious.

Low woods; Ashe, Avery Cos., NC; GA; GSMNP.

FIGURE 20	**4. *Carya pallida*** sand hickory

A small to medium-sized tree. Older bark light gray to dark gray, with a network of ridges and furrows.

Leaflets mostly seven (seven to nine), lanceolate to obovate or elliptical, 4–16 cm long, finely toothed, long pointed at apex, wedge shaped to rounded at base, light green and lustrous above, nearly glabrous beneath except for tufts of hairs on principal veins; *petioles yellow dotted or silvery dotted,* as is lower surface of leaflets.

Twigs slender, reddish brown, usually glabrous but *covered with golden dotlike scales* near tip.

Terminal bud *very small* for a hickory, 5–7 mm long, ovoid, pointed or blunt, reddish brown, *the outer scales dotted with yellowish and silvery scaly particles,* but outer scales commonly falling off early, exposing greenish or yellowish scales beneath. Lateral buds smaller and with fewer scales.

Fruits 2–3 cm long, including rather thin husk, ellipsoid to subglobose, yellow-scaly, thin shelled, kernel sweet.

Dry woods; occasional, essentially throughout.

FIGURE 21   Pignut hickory, *Carya glabra*. Leaf, twig, enlarged terminal bud, fruit, and silhouette of mature tree.

	5. *Carya glabra*
FIGURE 21	pignut hickory

A medium-sized to large tree. Older bark gray, usually tight, occasionally scaly or shaggy, with network of furrows and ridges.

Leaflets five to seven, lanceolate, ovate, or obovate, 4–22 cm long, finely toothed, short pointed or long pointed at apex, wedge shaped or rounded at base, dark green to yellowish green and glabrous above, paler and nearly glabrous beneath except for scattered hairs on lower veins.

Twigs relatively *slender,* reddish brown, scurfy-hairy when young, *glabrous* at maturity, often yellow dotted toward tip.

Terminal bud *small* for a hickory, 8–11 mm long, ovoid to subglobose. Bud scales about four visible, overlapping, the outer scales reddish brown to dark brown, glabrous or minutely hairy, keeled and extended into a slender point, often dotted more or less with yellowish scaly particles; but outer scales usually shed early, exposing lighter-colored inner scales, softly hairy. Lateral buds smaller, *widely divergent.*

Fruits variable in size and shape, mostly 2–3 cm long, including smooth or slightly winged husk, obovoid or subglobose, thick shelled, kernel variable in flavor, sweet or bitter.

Dry to moist woods at lower elevations; common, throughout.

Birch Family (Betulaceae)

Trees or shrubs; leaves alternate, deciduous, simple, usually double toothed; both staminate and pistillate flowers in catkins except in *Corylus,* which has pistillate flowers in few-flowered heads; fruit a nutlet, winged nutlet, or nut enclosed in a husk; birch, alder, hornbeam, hophornbeam, hazelnut.

Betula (birch)

Small to large trees. Bark ranging from white to brownish black, often scaling off in papery layers, and *banded horizontally with many elongated lenticels.*

Leaves alternate (often appearing opposite on short spurs), deciduous, simple, straight veined, single toothed or double toothed.

Twigs slender, usually somewhat zigzag, circular in cross section; *short spur twigs abundant on older growth, regularly bearing two leaves, or crowded leaf scars, and a terminal bud.* Lenticels dotlike on young twigs but elongating horizontally as twigs enlarge. Pith very small, green, clearly or obscurely three-angled in cross section.

Terminal buds present on short spurs, absent on long twigs. Buds solitary, ovoid, pointed, stalkless but commonly appearing stalked at tip of spur twigs. Visible bud scales about three on lateral buds, five to eight on terminal buds.

Leaf scars small, crescent shaped, triangular, or half elliptical. Bundle scars three. Stipule scars present, narrow.

Staminate catkins *exposed all winter* on mature trees, flowering in spring. Pistillate catkins hidden in buds until spring.

Fruits small, winged nutlets borne in a scaly, conelike structure, the scales usually deciduous at maturity, releasing nutlets.

1 Mature bark not white but yellowish gray, reddish and pinkish, or dark reddish brown to almost black:
  2 Buds small, mostly 4–6 mm long, flattened and appressed; leaves broadly wedge shaped or straight across at base; bark reddish brown to light pink, peeling and curling into ragged strips, not aromatic .... (1) *B. nigra*
  2' Buds larger, 5–10 mm long, not flattened and appressed; leaves rounded to heart shaped at base; bark aromatic with odor of wintergreen:
    3 Bark silvery or bronze colored, peeling into thin curls and ribbons; young twigs with yellowish tint, mildly aromatic..... (2) *B. alleghaniensis*
    3' Bark dark reddish brown to nearly black, close fitting, resembling cherry bark; young twigs dark brown or reddish brown, strongly aromatic ....(3) *B. lenta*

1′ Mature bark white or grayish white; rare and local:
　4 Buds 5–9 mm long; bark on young stems easily peeled off in thin layers;
　　leaves more or less heart shaped at base . . . .(4) *B. papyrifera*
　4′ Buds less than 5 mm long; bark on young stems tight, not easily peeled off;
　　leaves broadly wedge shaped to nearly straight across at base
　　. . . . . (5) *B. populifolia*

FIGURE 22 | **1. *Betula nigra***
　　　　　 | river birch

Bark on young trunks and branches *reddish brown or pink, peeling and curling, shaggy* as if partly torn away; old trunks scaly plated.

Leaves ovate, 4–9 cm long, sharply and mostly doubly toothed, short pointed at apex, very *broadly wedge shaped* to nearly straight across at base, green and glabrous above, pale and hairy to nearly glabrous beneath; petioles resin dotted and usually hairy.

Twigs brown to dark red, usually hairy, covered with many resin dots, *not aromatic;* bark soon becoming scaly and curly; lateral spurs numerous on older growth.

Buds *small,* usually 4–6 mm long, ovoid, pointed, more or less *flattened and pressed against twig,* reddish brown, hairy; bud scales five to seven visible on terminal buds, three on lateral buds.

On bottomlands and along streams; occasional, throughout.

FIGURE 22 | **2. *Betula alleghaniensis***
　　　　　 | yellow birch

Bark on young trunks and branches *yellowish or silvery, peeling horizontally into thin, curly, papery strips;* becoming rough and separated into platelike scales on old trunks.

Leaves ovate or ovate-oblong, 4–10 cm long, finely and sharply double toothed, long pointed at apex, slightly heart shaped, rounded, or unequal at base, dull green above, pale green and glabrous beneath except for whitish hairs along veins.

Twigs *yellowish brown,* brown, or greenish and hairy when young, aging to brown and becoming glabrous, with only *faint wintergreen odor and taste;* sparsely dotted with pale lenticels; spur twigs common on older growth.

Buds 5–10 mm long, ovoid, sharply pointed, reddish brown, mostly glabrous but scales ciliate; terminal buds present on short spurs, absent on long twigs; bud scales seven or eight on terminal buds, about three on lateral buds.

Moist woods; common at higher elevations, especially above 3,500 feet, throughout.

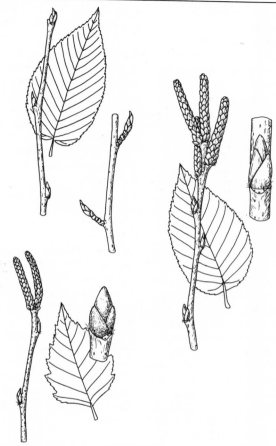

FIGURE 22   *Betula*. Upper left: sweet birch, *B. lenta,* twig, leaf, and older twig with
characteristic spur growth. Right: yellow birch, *B. alleghaniensis*, leaf, twig
with catkins, and enlarged lateral bud. Lower left: river birch, *B. nigra,* twig
with catkins, leaf, and enlarged uppermost lateral bud.

FIGURE 22	3. *Betula lenta* sweet birch, black birch, cherry birch

Bark of young trunks and branches *reddish brown to brownish black, not peeling
into thin layers,* resembling cherry bark, hence common name cherry birch;
eventually broken into scaly plates on old trunks.

Leaves ovate or elliptical, 5–12 cm long, sharply and doubly toothed, long pointed
at apex, usually heart shaped at base, green and generally lustrous above, hairy
only on veins beneath.

Twigs lustrous, light brown to *dark brown or reddish brown,* glabrous, with *strong
wintergreen odor and taste;* lenticels appearing as raised whitish dots elongating
horizontally with age; spur twigs numerous on older growth.

Buds somewhat divergent, 5–10 mm long, ovoid, sharp pointed, reddish brown,
glabrous or nearly so; terminal buds present on spur twigs, absent on longer
twigs; visible bud scales six or seven on terminal buds, about three on lateral
buds.

May be confused with *Carpinus* and *Ostrya,* which also have finely and doubly
toothed leaves, but this species can be distinguished by odor and flavor of twigs.

Rich woods; common below 4,000 feet, throughout.

	4. *Betula papyrifera*
FIGURE 23	paper birch

Bark at first *reddish brown* or brown, soon curling off and exposing creamy or
pinkish white layers beneath. *Older bark chalky white, peeling into loose curls,*
marked by elongated lenticels. Mature trunks black and scaly at base.

Leaves ovate to elliptical oblong, 5–10 cm long, irregularly double toothed, short
pointed or long pointed at apex, heart shaped at base (ours, var. *cordifolia*), dark
green above, paler and hairy at juncture of veins beneath.

Twigs reddish brown to orange brown, glabrous or slightly hairy, not aromatic;
lenticels conspicuous; many short spurs on older growth.

Buds 5–9 mm long, ovoid, pointed; bud scales hairy along margin, otherwise glabrous; three or four scales on lateral buds, about seven on terminal buds.

High mountains; Buncombe, Haywood, Macon, Yancey Cos., NC.

	5. *Betula populifolia*
	gray birch

Bark at first brownish, soon *grayish white, closely fitted,* not readily peeling; *inverted V-shaped markings* clearly visible at base of branches.

Leaves triangular in outline, 3–8 cm long, double toothed, long pointed at apex,
straight across, unequal, or broadly wedge shaped at base, dark shining green
and glandular roughened above, slightly paler and glabrous beneath.

Twigs reddish brown or orange brown to grayish, without hairs but liberally *covered with pimplelike resin dots.*

Buds small, mostly 3–4 mm long, ovoid, pointed, glabrous, often slightly resinous.
Bud scales at least three, overlapping.

Low woods; Avery and Macon Cos., NC.

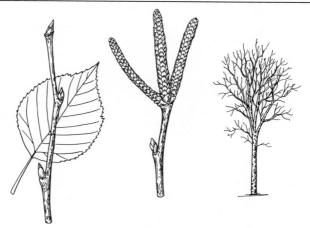

FIGURE 23   Paper birch, *Betula papyrifera*. Leaf, twig, twig with catkins, silhouette of mature tree.

Alnus (alder)

Deciduous shrubs or rarely small trees.

Leaves alternate, simple, pinnately veined, toothed.

Twigs more or less three-angled; pith small, three-sided in cross section.

Buds solitary, rather large, stalked or stalkless, with valvate or overlapping scales.

Both kinds of flowers in *catkins,* the pollen-producing (staminate) exposed in winter; seed-producing (pistillate) catkins exposed in winter in *A. serrulata,* appearing with the leaves on spring growth in *A. crispa.*

Leaf scars alternate, raised. Bundle scars three. Stipule scars present, narrow.

Fruit a small, *woody, conelike structure,* in clusters, persistent on plant long after seeds (nutlets) are shed.

1 Leaves mostly wedge shaped at base; buds clearly stalked; bud scales two (occasionally three), meeting at edges but not overlapping (valvate); both pistillate and staminate catkins naked in winter . . . . (1) *A. serrulata*
1' Leaves rounded or heart shaped at base; buds without stalks or merely short stalked; bud scales at least three, overlapping; pistillate catkins enclosed in scaly buds in winter . . . .(2) *A. crispa*

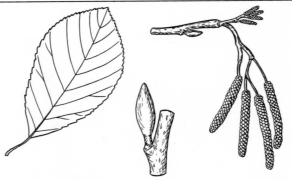

FIGURE 24   Hazel alder, *Alnus serrulata*. Leaf, enlarged bud, twig with both pistillate (female) and staminate (male) catkins.

| FIGURE 24 | 1. *Alnus serrulata*
hazel alder |

A thicket-forming shrub or rarely small tree.

Leaves obovate to elliptical, 5–13 cm long, with nearly straight veins, finely and sharply toothed, mostly blunt or rounded at apex, broadly wedge shaped to rounded at base, dark green and glabrous above, paler and glabrous or hairy along veins beneath.

Twigs slender, greenish gray to purplish brown, hairy to nearly glabrous.

Leaf buds *prominently stalked,* 6–9 mm long, including stalk, rounded or bluntly pointed at tip, dark red, often tinged with purple, more or less hairy, especially on stalk, covered by two (occasionally three) bud scales, all about as long as the bud.

Next season's staminate and pistillate *catkins naked in winter* on preceding year's growth, the staminate larger and more conspicuous than pistillate.

Leaf scars raised, three-sided to nearly circular. Bundle scars three, but often divided.

Woody fruit cones *short stalked,* in clusters of four to ten, long persistent after nutlets are shed.

Along streams and in wet woods; common, throughout.

| FIGURE 25 | 2. *Alnus crispa*
mountain alder, green alder |

A bushy shrub.

Leaves broadly ovate, 3–8 cm long, finely and sharply toothed, often irregularly or doubly toothed, the margin frequently puckered or ruffled, blunt to rounded at

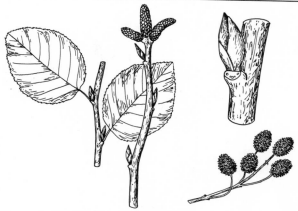

FIGURE 25   Mountain alder, *Alnus crispa*. Left to right: twig with leaves and buds, twig showing staminate (male) catkins, enlarged leaf bud and leaf scar, cluster of woody fruit "cones."

apex, rounded to somewhat heart shaped at base, green on both surfaces, more or less gummy and aromatic when young.

Twigs moderate in size, reddish brown to purplish, glabrous or nearly so (sometimes hairy near tip).

Buds short stalked or *nearly stalkless, long pointed* at tip, dark red to purplish, gummy; bud scales three or more, *overlapping.*

Next season's *pistillate catkins hidden in buds* and thus not evident in winter.

Leaf scars half circular to triangular.

Fruit cones *long stalked,* at least some stalks as long as or longer than the cones. Roan Mountain, Mitchell Co., NC.

	Carpinus caroliniana
FIGURE 26	American hornbeam, ironwood

A small tree. Trunk and larger branches *sinewy* or *fluted,* with low broad ridges, appearing somewhat like taut muscles. Bark *gray, tight, smooth.*

Leaves alternate, two-ranked, deciduous, simple, ovate or broadly lanceolate, mostly 4–10 cm long, sharply double toothed, principal vein extending to each large tooth, long pointed at apex, usually rounded at base.

Twigs *very slender,* circular in cross section, somewhat zigzag, dark brown or dark reddish brown, glabrous or hairy.

Terminal bud absent. Lateral buds *four-angled,* pointed, with eight to twelve, occasionally more, visible scales, and of *two distinct sizes*—smaller leaf buds 2–

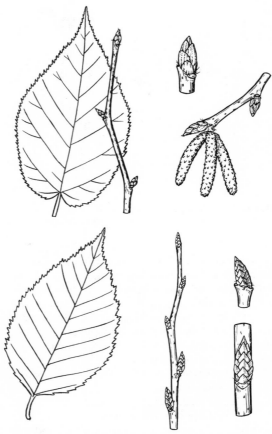

FIGURE 26 Above: eastern hophornbeam, *Ostrya virginiana*. Leaf, twig, enlarged bud at end of twig showing leaf scar and twig scar, enlarged twig with catkins. Below: American hornbeam, *Carpinus caroliniana*. Leaf, twig, enlarged leaf bud at end of twig, enlarged flower bud and leaf scar.

3 mm long, and larger, plumper flower buds 3–5 mm long. Bud scales mostly glabrous, faintly striated, reddish brown.

Leaf scars very small, raised, flattened-elliptical to crescent shaped. Bundle scars three, often indistinct. Stipule scars present.

Staminate and pistillate catkins enclosed in scaly buds in winter and therefore not visible.

Fruit a small nutlet borne at base of a three-pointed leafy bract, in loose, drooping
    clusters.
This species is apt to be mistaken for *Ostrya* when young, but the buds are smaller,
    usually four-sided, and have more scales visible.
Bottomlands and rich woods; common, throughout.

FIGURE 26	*Ostrya virginiana* eastern hophornbeam

A small, slender tree. Bark of trunk *thin,* grayish brown, *in narrow vertical strips,*
    these often loose at the ends, sometimes twisted.
Leaves alternate, two-ranked, deciduous, simple, ovate to elliptical, 2–15 cm long,
    sharply and usually double toothed (resembling leaves of elm or birch), long
    pointed at apex, heart shaped to rounded at base, more or less hairy beneath, at
    least along larger veins and in vein axils; lateral veins often somewhat curved
    and branched near margin.
Twigs slender, zigzag, reddish brown, at first hairy with long, straight hairs, later
    nearly glabrous.
Buds all nearly alike in shape and number of scales, ovoid, 2–6 mm long, pointed,
    circular in cross section, *not angled* as in *Carpinus;* visible bud scales six or
    more, greenish or brownish, *striated* with fine, parallel vertical grooves (under
    magnifying lens). Terminal bud absent.
Staminate catkins *exposed in winter,* usually in twos or threes, on mature trees.
Leaf scars two-ranked, small, half circular to elliptical. Bundle scars three. Stipule
    scars present, narrow.
Fruit a hoplike cluster or bladdery sacs, each enclosing a nutlet; ripening in late
    summer or autumn, usually falling before winter.
Moist or dry woods; infrequent, essentially throughout.

Corylus (hazelnut)

Shrubs, often forming colonies.
Leaves alternate, apparently two-ranked, at least on lateral twigs, deciduous,
simple, irregularly single toothed or *double toothed.*
Twigs slender to moderate, zigzag.
Terminal bud absent. Lateral buds rounded or broadly ovoid, blunt, with four
to six scales visible.
Staminate flowers in catkins appearing in autumn, *conspicuous all winter,* ma-
turing in following spring. Pistillate flowers in scaly buds until spring.
Leaf scars half circular to triangular. Bundle scars three or more, often obscure
and difficult to see. Stipule scars present, unequal in size.
Fruit a nut in close-fitting husk made up of two enlarged, leafy bracts; nuts sweet
and edible.

1　Twigs and leaf petioles usually covered with gland-tipped hairs; staminate catkins with stalk 3–6 mm long; beak of fruit 1–2 cm long
. . . . (1) *C. americana*

1′　Twigs and leaf petioles without gland-tipped hairs except near nodes; staminate catkins stalkless or nearly so; beak of fruit 3–4 cm long. . . . . (2) *C. cornuta*

| | 1. *Corylus americana* |
| FIGURE 27 | American hazelnut |

Leaves ovate to nearly circular, 6–15 cm long, double toothed, short pointed or long pointed at apex, heart shaped or slightly unequal at base, dull dark green and glabrous above, paler and finely hairy beneath, at least along veins; petioles hairy, some hairs *gland tipped*.

Twigs brown or reddish brown; young twigs *bristly with glandular hairs* (may be sparse late in season).

Buds ovoid, 2–4 mm long, blunt, reddish or purplish, somewhat hairy, especially near tip, with four to six overlapping scales visible.

Staminate catkins slender, *short stalked*.

Fruit husk *not much longer than the fruit*.

Rich woods, thickets, roadsides, clearings; infrequent, throughout.

| | 2. *Corylus cornuta* |
| FIGURE 27 | beaked hazelnut |

Leaves ovate, 2–10 cm long, double toothed or irregularly single toothed, long pointed at apex, heart shaped or rounded at base, glabrous or hairy above, hairy beneath; petioles hairy but not glandular.

Twigs reddish brown, glabrous or with a few pale hairs near nodes and toward tip of twig; *not glandular-bristly*.

Buds ovoid, 2–5 mm long, reddish brown, with four to six scales showing, outer pair of scales somewhat hairy to nearly glabrous, *soon deciduous,* the inner scales gray-silky.

Staminate catkins *without stalks* or very short stalked.

Fruit husk *extended into a slender, tubelike beak,* 3–4 cm beyond fruit.

Thickets and edges of woods; infrequent, throughout.

FIGURE 27   *Corylus*. Above: American hazelnut, *C. americana*. Bud, leaf, twig with catkins, fruit. Below: beaked hazelnut, *C. cornuta*. Bud, leaf, twig with catkins, fruit.

Beech Family (Fagaceae)

Trees or occasionally shrubs; leaves alternate, simple, in our species deciduous, usually toothed or lobed; flowers small, typically unisexual, the staminate in long, narrow catkins, the pistillate in spikes, both forms on same plant; fruit a nut or an acorn; beech, chestnut, oak.

	Fagus grandifolia
FIGURE 28	American beech

A large tree with *smooth, light gray to dark gray bark,* even on old trunks.

Leaves alternate, two-ranked, deciduous, simple, elliptical to ovate, 6–15 cm long, *coarsely toothed with sharp teeth,* long pointed at apex, wedge shaped to slightly heart shaped at base; lateral veins straight and parallel, main vein extending into each tooth; dark bluish green above, yellowish green and lustrous beneath.

Twigs slender, moderately zigzag, reddish brown to brown, glabrous or nearly so; *nearly encircled at nodes by narrow stipule scars;* slow-growing twigs numerous; base of each year's growth marked by many closely set scars left by fallen bud scales.

Terminal bud present, *conspicuously long and slender,* 10–25 mm long, *sharp pointed,* reddish brown or brown, lustrous, with ten or more overlapping scales visible. Lateral buds similar, usually a little smaller, widely divergent, tapering to a narrow base but not stalked.

Leaf scars small, raised, half circular to elliptical. Bundle scars three, or the middle one divided. Stipule scars long and narrow, extending well around twig but not quite meeting.

Fruit a three-cornered, pale brown, shiny nut, usually borne in pairs, enclosed in a *prickly bur;* nut sweet and edible.

Cool, fertile woods; common at higher elevations, throughout.

Castanea (chestnut, chinkapin)

Trees or shrubs.

Leaves alternate, mostly two-ranked, deciduous, simple, coarsely toothed, the veins straight or nearly so and extending to margin, *the teeth bristle tipped.*

Twigs slender to moderately stout, more or less fluted below nodes, usually brownish, glabrous or woolly. *Pith five-sided in cross section.*

Terminal bud absent; lateral buds *blunt,* light brown to dark brown, with *two or three scales showing.*

Leaf scars raised, half circular, elliptical, or nearly circular. Bundle scars three or sometimes more. Stipule scars present, narrow, unequal in size.

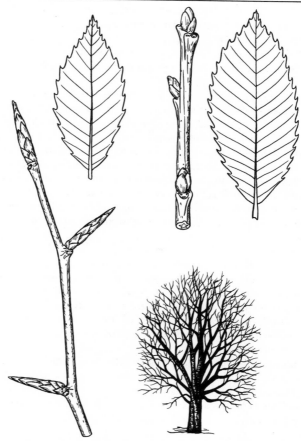

FIGURE 28   Above left and below: American beech, *Fagus grandifolia*. Leaf, twig with buds and leaf scars, silhouette of mature tree. Above right: American chestnut, *Castanea dentata*. Twig showing buds, leaf scars, and twig scar at end of twig; leaf.

Fruit a large, rounded nut, solitary or two or three together, *enclosed in a prickly bur.*

1 Leaves mostly more than 15 cm long, green and glabrous beneath; twigs and buds glabrous or essentially so .... (1) *C. dentata*
1' Leaves mostly less than 15 cm long, white-woolly beneath; twigs and buds hairy .... (2) *C. pumila*

	1. *Castanea dentata*
FIGURE 28	American chestnut

A large tree, formerly abundant in the mountains, now mostly destroyed by chestnut blight, a bark disease; young plants are frequently seen growing from old roots.

Leaves oblong-lanceolate, 13–25 cm long (usually more than 15 cm), straight veined, coarsely toothed, the teeth curved forward and bristle tipped; green and *glabrous* on both sides; leaves similar to those of beech (*Fagus grandifolia*) but longer, wider, and with sharper teeth.

Twigs slightly zigzag, reddish brown to greenish brown or sometimes purplish, nearly glabrous; lenticels numerous, white; pith five-sided in cross section.

Buds ovoid, 4–6 mm long, somewhat resembling grains of wheat, blunt, brown, glabrous; visible scales two or three. Terminal bud commonly absent.

Fruits spiny burs, each enclosing two or three flattened nuts; seed sweet, edible.

Rich woods; occasional, mostly stump sprouts, throughout.

	2. *Castanea pumila*
FIGURE 29	Allegheny chinkapin

A large shrub or small tree, not so susceptible to blight as *C. dentata.*

Leaves elliptical to obovate, 6–20 cm long (mostly 7–15 cm), straight veined, sharply toothed with bristle-tipped teeth, pointed at apex, wedge shaped to rounded at base, dark green and glabrous above, *whitish beneath with a velvety covering of hairs.*

Twigs at first woolly-hairy, later nearly glabrous, reddish brown to purplish brown.

Buds ovoid, mostly 2–3 mm long, short-hairy, reddish or purplish, with two or three scales exposed.

Fruits spiny burs, each enclosing *only one nut;* seed sweet and edible.

Dry woods and thickets; fairly common, NC, SC, GA.

Quercus (oak)

Trees or rarely shrubs.

Leaves alternate, in our species deciduous, but dry leaves often persisting on tree into winter; simple, pinnately veined, entire, lobed, or variously toothed, commonly quite variable on same tree.

Twigs slender to stout, straight, rigid, more or less fluted; branching in distinctive pattern, several twigs radiating from ends of previous year's growth. Pith *five-angled* in cross section.

Buds *three or more in a cluster at twig tips.* Terminal bud present, about same size as lateral buds in terminal cluster. Lower laterals similar but smaller. Buds covered by *many scales in five nearly vertical rows.*

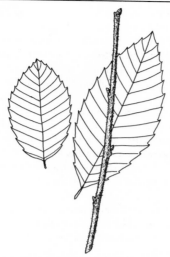

FIGURE 29   Allegheny chinkapin, *Castanea pumila*. Two forms of leaves, twig.

Leaf scars raised, tending to be crowded toward end of twig. Bundle scars *numerous*, scattered. Stipule scars present, small.

Fruit an *acorn*, partly enclosed by a scaly *cup*.

Summer Key

1 Leaf apex, lobes, or teeth bristle tipped (bristles are an extension of the principal veins); acorns maturing in second year; seeds bitter, not edible; bark of trunk usually dark and furrowed (black oaks and their relatives):

  2 Leaves entire, except for bristle at apex; upper surface glossy, lower surface softly hairy .... (1) *Q. imbricaria*

  2′ Leaves all or mostly lobed:

    3 Leaves uniformly hairy or scurfy beneath:

      4 First-year twigs glabrous, often shiny late in season; leaves broadest at or slightly above middle, deeply lobed throughout length, sometimes nearly glabrous.............................(2) *Q. velutina*

      4′ First-year twigs scurfy-hairy, at least near tip:

        5 Leaves broadest toward apex, all or mostly three-lobed:

          6 Leaves coated with yellowish or grayish hairs beneath; mature buds at twig tip 3–8 mm long, not or only slightly angled

          ....(3) *Q. falcata*

          6′ Leaves coated with rusty brownish hairs beneath; mature buds at twig tip 6–12 mm long, strongly angled

          ....(4) *Q. marilandica*

        5′ Leaves broadest near middle, mostly five-lobed, the middle lobe

　　　　　usually long and narrow, often curved to left or right, and often
　　　　　with several toothlike lobes near outer end............(3) *Q. falcata*
　　3′ Leaves glabrous beneath, or scurfy-hairy when young but soon becom-
　　　　ing glabrous, or hairy only on midrib or in axils of veins:
　　　　　7 Leaves five- to eleven-lobed less than halfway to midrib, or if lobed
　　　　　　more than halfway, the middle unlobed part of blade always more
　　　　　　than 2 cm across:
　　　　　　8 Buds strongly angled, densely hairy, yellowish gray; leaves glossy
　　　　　　　green above .... (2) *Q. velutina*
　　　　　　8′ Buds not strongly angled, glabrous or hairy only near tip, dark
　　　　　　　red or reddish brown; leaves dull green above ....(5) *Q. rubra*
　　　　　7′ Leaves deeply seven-lobed (five to nine lobes) nearly to midrib, the
　　　　　　middle unlobed part of blade often less than 2 cm across
　　　　　　...(6) *Q. coccinea*
1′ Leaf apex and lobes or teeth never bristle tipped; acorns maturing in first
　year; seeds sweetish, edible; bark of trunk usually light colored and scaly
　(dark in *Q. prinus*) (white oaks and their relatives):
　9 Leaves lobed (white oaks):
　　10 Leaves thick and somewhat leathery, typically lobed so as to resemble
　　　a cross, rough on upper surface, finely hairy beneath when mature
　　　... (7) *Q. stellata*
　　10′ Leaves thinner in texture, rather evenly and usually deeply seven- to
　　　eleven-lobed, smooth on upper surface, glabrous and pale or whitish
　　　beneath when mature ....(8) *Q. alba*
　9′ Leaves coarsely toothed, not lobed; lateral veins nearly straight and par-
　　allel, extending to teeth (chestnut oaks):
　　11 Leaves mostly 5–8 cm long, with three to eight veins and teeth on
　　　each side, the teeth blunt or slightly pointed .... (9) *Q. prinoides*
　　11′ Leaves more than 8 cm long, with eight to sixteen veins and teeth on
　　　each side:
　　　12 Teeth of leaves usually blunt or rounded; bark dark and deeply
　　　　furrowed; one of our most common oaks ....(10) *Q. prinus*
　　　12′ Teeth of leaves fairly sharp, each tipped with a gland; bark light
　　　　gray and scaly; rare in the mountains
　　　　....................................... (11) *Q. muehlenbergii*

Winter Key
1 Larger buds in terminal cluster 5 mm or more in length:
　2 Buds hairy, often densely so, over entire surface or only on upper half:
　　3 Buds uniformly hairy or woolly, often more than 9 mm long:
　　　4 Twigs scurfy or hairy; buds rusty brown............(4) *Q. marilandica*
　　　4′ Twigs glabrous, often shiny; buds yellowish gray to grayish brown
　　　... (2) *Q. velutina*

3' Buds hairy mostly above middle, rarely more than 9 mm long:
   5 Buds sharply pointed; twigs usually scurfy-hairy toward tip
     . . . .(3) *Q. falcata*
   5' Buds blunt or rounded; twigs glabrous. . . . . (6) *Q. coccinea*
2' Buds glabrous, or minutely hairy, or hairy only at tip:
   6 Buds slender, conical, sharply pointed, the scales chestnut brown, not
     glossy; bark on old trunks with deep, sharply angled furrows and
     ridges . . . .(10) *Q. prinus*
   6' Buds more plump, ovoid, sharply to bluntly pointed, the scales dark
     reddish brown or dark red, usually glossy; bark on old trunks with
     long, smooth plates between furrows . . . .(5) *Q. rubra*
1' Larger buds in terminal cluster mostly less than 5 mm long:
   7 Twigs scurfy-woolly or covered with matted hairs, at least toward tip:
     8 Buds plump, not much longer than broad, rounded or merely acute at
       tip; lateral buds usually set at less than forty-five-degree angle to twig
       . . . . (7) *Q. stellata*
     8' Buds rather slender, considerably longer than broad, pointed at tip;
       lateral buds widely divergent, often at nearly right angle to twig
       . . . .(3) *Q. falcata*
   7' Twigs glabrous or nearly so:
     9 Buds whitish-silky in upper half, the lower half glabrous and reddish
       (a few buds may be hairy almost to base) . . . . (6) *Q. coccinea*
     9' Buds glabrous or essentially so:
      10 Buds rounded or bluntly pointed:
        11 Twigs moderately stout, usually 2.5 mm or more across,
          tinged with red or purple, often with whitish bloom and ap-
          pearing polished or pearly . . . .(8) *Q. alba*
        11' Twigs slender, less than 2.5 mm across, brown to dark gray
          . . . . (9) *Q. prinoides*
      10' Buds sharply pointed:
        12 Buds dull, circular or nearly so in cross section, the scales
          pale or straw colored along edges; twigs orange brown or
          straw brown; old bark light colored and flaky
          . . . .(11) *Q. muehlenbergii*
        12' Buds glossy, somewhat angled in cross section, the scales
          hairy or jagged along edges; twigs reddish brown or greenish
          brown to grayish green; old bark blackish and rough
          . . . . (1) *Q. imbricaria*

Except for dwarf chinkapin oak (*Q. prinoides*), all the oaks described in this
book reach tree size. A few species, especially in the black oak group, are so variable
in the form and size of their leaves that they often puzzle even trained botanists.
Similarly, the winter traits of some species are variable, and you may come upon

FIGURE 30   Shingle oak, *Quercus imbricaria*. Leaf, fruit, and enlarged clustered buds at
end of twig.

individual plants that are difficult to key out properly. Keeping all this in mind, the
beginner should not be disappointed with occasional failures. Skill in identifying
the oaks comes with experience and effort.

FIGURE 30	**1. *Quercus imbricaria*** shingle oak

A medium-sized, graceful tree. Old bark blackish and rough.

Leaves lanceolate, elliptical, or oblong, mostly 8–15 cm long, *entire,* short pointed
to rounded and bristle tipped at apex, rounded to wedge shaped at base, dark
green, glossy, and glabrous above, pale green or brownish and softly hairy
beneath.

Twigs slender, greenish gray to reddish green, glabrous, lustrous.

Buds in terminal cluster mostly 2–4 mm long, ovoid, sharply pointed, *angled,* light
brown, *lustrous,* the scales ragged or hairy along edges.

Acorns 12–15 mm long; cup deeply saucer shaped; cup scales blunt, hairy.

In moist woods, along streams, and on hillsides; rare, NC, GSMNP.

FIGURE 31	**2. *Quercus velutina*** black oak

A large tree. Old bark nearly or quite *black and very rough;* inner bark *orange
yellow.*

Leaves widely obovate to ovate in general outline, 13–20 cm long, more or less
deeply lobed, the lobes five to nine but usually seven and often toothed; green
and *lustrous* above, green or yellowish green and somewhat scurfy-hairy beneath or often completely glabrous late in season.

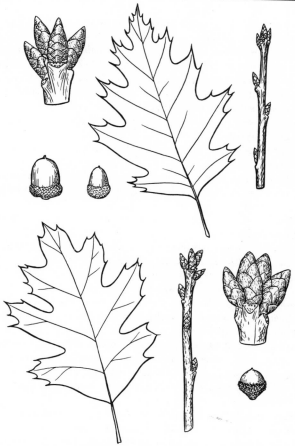

FIGURE 31   *Quercus.* Above: northern red oak, *Q. rubra.* Enlarged terminal bud cluster, two forms of fruit, leaf, and twig. Below: black oak, *Q. velutina.* Leaf, twig, enlarged terminal bud cluster, and fruit.

Twigs stout, prominently fluted, reddish brown, glabrous, or sometimes scurfy at first but soon becoming glabrous.

Buds in terminal cluster *large,* 5–13 mm long, ovoid or conical, sharply pointed, *strongly angled* or grooved, *densely woolly* with yellowish gray hairs.

Acorns 15–20 mm long; cup deep, bowl shaped, covering about half of nut; cup scales loosely overlapping, forming a loose fringe at the rim.

Upland woods; common, throughout.

	3. *Quercus falcata*
FIGURE 32	southern red oak, Spanish oak

A large tree. Mature bark dark gray, with shallow furrows and broad ridges.

Leaves obovate to ovate in outline, mostly 10–20 cm long, and either (*a*) shallowly to rather deeply three-lobed at apex, often similar to leaves of *Q. marilandica*, or (*b*) more or less deeply five- to seven-lobed, the lobes bristle tipped, the *central lobe usually long and narrow,* commonly with several toothlike lobes near end and often curved; shiny green and glabrous above, *woolly* with yellowish gray hairs beneath.

Twigs moderate to stout, dark red to reddish brown, scurfy-hairy or nearly glabrous.

Buds in terminal cluster 3–8 mm long, ovoid, pointed, more or less angled, chestnut brown and *rusty-hairy,* at least above middle. Lower lateral buds similar, *widely divergent.*

Acorns 10–15 mm long; cup deeply or shallowly saucer shaped; cup scales closely appressed, hairy.

Moist to dry woods; throughout, mostly at lower elevations.

	4. *Quercus marilandica*
FIGURE 32	blackjack oak

A small tree of poor or dry soils. Bark of trunk very dark, rough, and divided into *small, squarish blocks.*

Leaves thick and leathery, 10–25 cm long, broadly obovate, suggesting pear in outline, entire below middle, *shallowly three-lobed at outer end,* or sometimes scarcely lobed at all but with bristle tips usually showing, rounded or heart shaped at base, dark yellowish green and lustrous above, *brownish-scurfy beneath.*

Twigs stout, reddish brown, scurfy with star-shaped clusters of hairs.

Buds in terminal group 6–12 mm long, ovoid to broadly ellipsoid, slender pointed, distinctly angled, *rusty-woolly,* similar to buds of *Q. velutina* but more reddish brown.

Acorns about 2 cm long; cup deep, covering about half of nut; cup scales hairy, loosely appressed.

Dry open woods; infrequent, essentially throughout.

	5. *Quercus rubra*
FIGURE 31	northern red oak, red oak

A medium-sized tree. Bark of trunk dark gray or blackish, shallowly furrowed, *ridges usually wide, smooth, and shiny.*

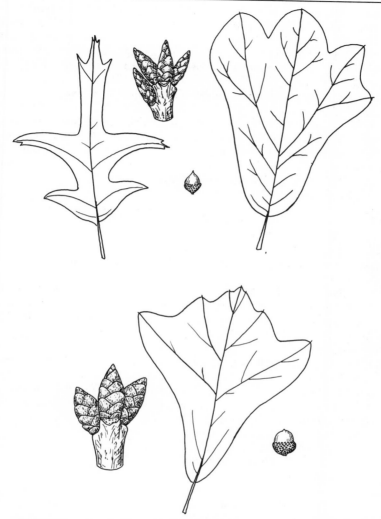

FIGURE 32   *Quercus*. Above: southern red oak, *Q. falcata*. Two forms of leaves, fruit, and enlarged terminal bud cluster. Below: blackjack oak, *Q. marilandica*. Enlarged terminal bud cluster, leaf, and fruit.

Leaves oblong to obovate in outline, 10–20 cm long, with seven to eleven bristle-tipped lobes, the lobes pointing forward and often toothed, separated by regular sinuses that extend *about halfway* to midrib; long pointed at apex, rounded or wedge shaped at base, *dull* dark green above, lighter green beneath, hairy when young but *nearly glabrous* at maturity, except for tufts of hairs in lower vein axils.

Twigs moderately stout, strongly fluted, reddish brown to greenish brown, glabrous, lustrous.

Buds in terminal cluster 5–8 mm long, narrowly ovoid, pointed, circular or slightly angled in cross section, *dark red or dark reddish brown, glabrous* or yellowish-hairy toward tip. Lateral buds divergent.

Acorns large, 2–3 cm long; cup shallow, enclosing only base of nut; cup scales thin, tightly appressed, hairy, their tips noticeably darkened.

Upland woods; common, throughout.

FIGURE 33 | 6. *Quercus coccinea*
scarlet oak

A large tree. Mature bark dark and rough, suggestive of *Q. velutina,* but *inner bark reddish.*

Leaves elliptical to ovate in outline, 7–20 cm long, *deeply divided into five to nine bristle-tipped lobes,* the unlobed part of blade commonly less than 2 cm wide, bright lustrous green above, a little paler and glabrous beneath except for tufts of hairs in axils of veins; turning scarlet in autumn, hence name.

Twigs slender, brownish red to purplish gray, lustrous.

Buds in terminal cluster 4–8 mm long, ovoid, blunt or rather sharply pointed, angled, *whitish-hairy above middle,* glabrous toward base.

Acorns 12–22 mm long, the nut usually with *several concentric rings around tip;* cup bowl shaped or top shaped, covering one-third to half of nut; cup scales glabrous, appressed, not spreading.

Dry sandy slopes and ridges; fairly common, throughout.

FIGURE 34 | 7. *Quercus stellata*
post oak

A medium-sized tree. Bark of trunk grayish, darker and more deeply furrowed than that of *Q. alba.*

Leaves *firm and leathery,* obovate in general outline, mostly 9–15 cm long, usually deeply five-lobed, the middle pair of lobes largest, *suggesting a cross,* green on both sides, shiny but rough above, *finely hairy beneath* with star-shaped clusters of hairs, velvety to touch.

Twigs stout, stiff, orange brown to reddish brown, *scurfy-woolly* with grayish or brownish hairs, at least toward tip.

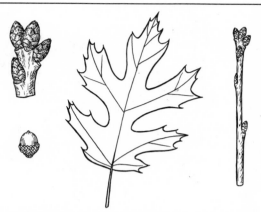

FIGURE 33   Scarlet oak, *Quercus coccinea*. Buds clustered at end of twig, fruit, leaf, and twig.

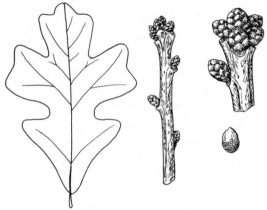

FIGURE 34   Post oak, *Quercus stellata*. Leaf, twig, fruit, and enlarged terminal bud cluster.

Buds in terminal cluster 2–5 mm long, bluntly pointed or rounded, circular (not angled) in cross section, light brown to reddish brown, short-hairy to nearly glabrous.

Acorns 12–20 mm long; cup *deep,* top shaped or bowl shaped, covering three-eighths to half of nut; cup scales small, thin, closely appressed, finely hairy.

Poor or sandy uplands; fairly common, throughout.

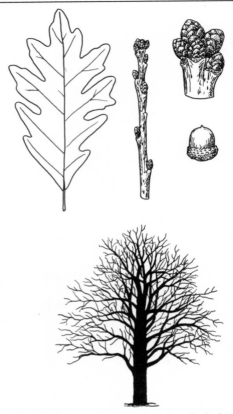

FIGURE 35   White oak, *Quercus alba*. Leaf, twig, enlarged buds clustered at end of twig, fruit, silhouette of mature tree.

	8. *Quercus alba*
FIGURE 35	white oak

A large tree. Bark of trunk thin, *light gray, shallowly furrowed and scaly.*

Leaves oblong-obovate in outline, 7–20 cm long, rather evenly divided into *seven to nine rounded lobes without bristle tips,* rounded at apex, wedge shaped at base, bright green above, glabrous and pale or whitened beneath.

Twigs moderately stout, strongly fluted, grayish red to greenish gray tinged with purple, glabrous, *smooth and polished.*

Buds in terminal cluster 3–5 mm long, *dome shaped* to broadly ovoid, bluntly pointed, reddish brown, glabrous.

FIGURE 36   Dwarf chinkapin oak, *Quercus prinoides*. Leaf, twig, and fruit.

Acorns 15–25 mm long; cup shallow, covering one-third or less of nut; cup scales
warty or knobby.
Coves, upland woods; common, throughout.

| | 9. *Quercus prinoides* |
| FIGURE 36 | dwarf chinkapin oak |

A shrub, rarely a small tree.
Leaves obovate to oblanceolate, 5–10 cm long, shallowly toothed with *three to
eight pointed or rounded teeth on each side,* blunt or pointed at apex, wedge
shaped at base, dark yellowish green and somewhat lustrous above, pale green
and usually white-woolly beneath.
Twigs slender, brittle, brown to dark gray, hairy when young, later glabrous.
Buds *small,* mostly 2–3 mm long, ovoid or conical-ovoid, blunt tipped or rounded,
pale brown, glabrous.
Acorns 10–15 mm long, ovoid; cup bowl shaped, enclosing one-third to half of nut;
cup scales small, hairy.
Dry woods and ridgetops; occasional, NC, SC, GA.

| | 10. *Quercus prinus* |
| FIGURE 37 | chestnut oak |

A medium-sized to large tree. Bark of trunk brown to black, with *prominent ridges
and deep, V-shaped furrows.*
Leaves elliptical to obovate, or sometimes nearly lanceolate, 10–20 cm long,
coarsely toothed with *broad rounded teeth* (in contrast to sharp, bristle-tipped
teeth of *Castanea*); abruptly long pointed at apex, mostly wedge shaped at base,
yellowish green and lustrous above, paler and often finely hairy beneath.
Twigs moderately stout, brown or reddish brown to greenish brown, glabrous.
Buds in terminal cluster 6–9 mm long, *distinctly conical,* sharply pointed, slightly
angled in cross section, light chestnut brown to light reddish brown, glabrous or

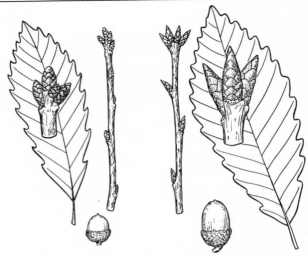

FIGURE 37   *Quercus.* Left: chinkapin oak, *Q. muehlenbergii.* Leaf, enlarged terminal
bud cluster, twig, fruit. Right: chestnut oak, *Q. prinus.* Twig, leaf, enlarged
terminal bud cluster, fruit.

slightly hairy; buds similar to those of *Q. rubra* but lighter colored. Lower lateral
buds widely divergent.

Acorns large, 2–3 cm long, lustrous; cup bowl shaped, covering one-third to half of
nut; cup scales more or less fused, free only at tips.

Dry upland woods and rocky ridges; common throughout, mostly below 4,000 feet.

FIGURE 37 | **11. *Quercus muehlenbergii***
chinkapin oak, chinquapin oak

A medium-sized to large tree. Old bark *light gray and flaky,* resembling that of
Q. alba.

Leaves lanceolate to narrowly oblong, 5–18 cm long, with eight to thirteen *sharp
or incurved teeth* on each side, pointed at apex, wedge shaped at base, yellowish
green and lustrous above, paler and whitish-hairy to glabrous beneath.

Twigs slender, brown to orange brown, glabrous.

Buds in terminal cluster 3–4 mm long, ovoid or conical, sharply or bluntly pointed,
same color as twigs, glabrous; similar to buds of *Q. prinus* but smaller.

Acorns 12–25 mm long; cup bowl shaped, enclosing about half of nut; cup scales
thin, long pointed, grayish brown, woolly.

Rocky hillsides, dry bluffs, generally on soils derived from limestone; rare, NC,
SC, GA.

Elm Family (Ulmaceae)

Small to large trees; leaves alternate in two rows, simple, in our species deciduous, mostly unequal at base; flowers small, inconspicuous, solitary or clustered; fruit in our species an evenly winged samara or a drupe; elm, sugarberry, hackberry.

Ulmus (elm)

Small to large trees.

Leaves alternate, *two-ranked,* deciduous, simple, straight veined, double toothed; the left and right sides of blade more or less *unequal at base,* the margin joining petiole at different points.

Twigs slender, slightly zigzag, circular in cross section.

Terminal bud absent. Lateral buds commonly appearing *tilted,* not pointing straight forward, and covered with scales overlapping in two vertical rows. Buds usually of two kinds: slender leaf buds near twig tips, and stout flower buds at lower positions.

Leaf scars two-ranked, half circular to elliptical. Bundle scars typically three, distinct, *sunken below surface of leaf scar.* Stipule scars present, small.

Fruit a *flattened,* roundish samara, *winged all around,* with centrally placed seed, maturing in spring (in our species).

1 First-year twigs hairy and rough to touch; buds rusty-hairy, at least near tip
....(1) *U. rubra*
1' First-year twigs glabrous or hairy but not rough to touch; buds not rusty-hairy:
  2 Leaves less than 4 cm wide, with petioles 1–3 mm long; twigs often with two opposite corky wings ....(3) *U. alata*
  2' Leaves more than 4 cm wide, with petioles 3–10 mm long; twigs without corky wings...(2) *U. americana*

FIGURE 38 | 1. *Ulmus rubra*
slippery elm

A medium-sized tree. Mature bark thick, rough, dark brown, with ridges more nearly parallel than in *U. americana;* bark ridges, when cut with knife, do not show whitish streaks.

Leaves ovate-oblong to elliptical, *large,* 9–15 cm long (occasionally to 20 cm), 5–9 cm wide, double toothed, abruptly short pointed to long pointed at apex, usually

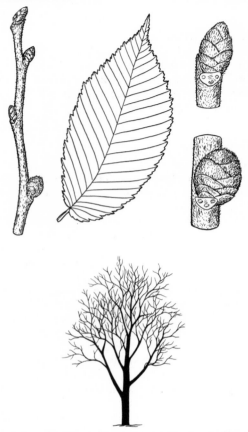

FIGURE 38   Slippery elm, *Ulmus rubra*. Twig with leaf buds and a flower bud, leaf, enlarged leaf bud, enlarged flower bud, silhouette of mature tree.

markedly asymmetrical at base, *very rough and sandpapery above,* hairy beneath, with cottony tufts of hairs in vein axils on lower side.

Twigs slender or moderately slender, ashy gray or brownish gray to reddish brown, hairy and *rough to touch; inner bark mucilaginous* when chewed.

Buds dark chestnut brown to nearly black, *rusty-hairy,* especially at tips, with six or more scales visible. Leaf buds narrowly ovoid, 5–6 mm long, bluntly pointed; flower buds larger and plumper than leaf buds, the scales two-ranked.

Fruits nearly circular, 12–19 mm long, flat, one-seeded, the wing light green, broad, glabrous, slightly notched at tip.

Moist woods at lower elevations; infrequent, throughout.

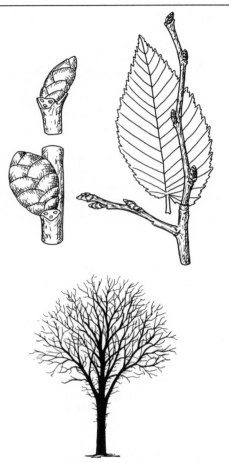

FIGURE 39   American elm, *Ulmus americana*. Enlarged topmost lateral bud, enlarged
              flower bud, both showing characteristic one-sided position of bud above leaf
              scar; leaf and twig; silhouette of mature tree.

| | 2. *Ulmus americana* |
| FIGURE 39 | American elm |

A large tree, typically *vase shaped* where open grown, the branches gracefully arching. Outer bark of trunk, when cut with knife, shows alternating reddish brown
  tissue and light colored streaks.
Leaves ovate-oblong to obovate, 10–15 cm long, 4–9 cm wide, coarsely double

toothed, long pointed at apex, decidedly unequal at base, dark green and smooth or rough on upper surface, paler and generally slightly hairy beneath, with tufts of hairs in vein axils; leaves resembling those of slippery elm but smaller, not so rough to touch, and more nearly glabrous on lower side.

Twigs slender, *dull brown* to reddish brown, glabrous or sparingly hairy, without corky wings.

Leaf buds narrowly ovoid, 3–6 mm long, somewhat pointed, flower buds 5–7 mm long, rounded, flattened. Bud scales reddish brown with darker borders, usually glabrous but minutely hairy along margin; buds often tilted a little to one side of leaf scar.

Fruits elliptical, 10–12 mm long, flat, one-seeded, the wing fringed with hairs but otherwise glabrous, deeply notched; in long-stemmed clusters.

Moist, rich woods at low elevations; Swain Co., NC; GA; GSMNP; highly susceptible to Dutch elm disease, a fungus spread by a beetle.

FIGURE 40 | 3. *Ulmus alata*
winged elm

A small or medium-sized tree. Mature bark light brownish gray, with irregular furrows and flat-topped ridges; if cut with knife, ridges show alternate patches of brown and yellow or pink.

Leaves *small,* narrowly elliptical, mostly 4–7 cm long and less than half as wide, coarsely double toothed, long pointed at apex, rounded at base, dark green, often lustrous, and smooth or slightly rough above, pale green and softly hairy beneath.

Twigs *very slender,* green to reddish brown, hairy or glabrous; lenticels small, raised, orange colored; commonly with *two opposite corky wings.*

Leaf buds narrowly ovoid, 2–3 mm long, flattened laterally, pointed, the scales reddish brown, somewhat scurfy, the margin dark and often with a few hairs; flower buds more plump, about 4 mm long, reddish brown; buds often not directly above leaf scar.

Fruits elliptical, about 10 mm long, reddish, flat, one-seeded, the wing narrow.

Low woods and streambanks; occasional, throughout.

Celtis (hackberry)

Small or medium-sized trees (our species). Bark of trunk gray to light brown, mostly smooth but with prominent corky knobs or roughened by steep-sided ridges, some appearing to stand on edge.

Leaves alternate, two-ranked, deciduous, simple, *palmately veined,* rounded and usually *unequal sided at base,* entire or toothed; branching two-ranked also, the lateral twigs lying all in one plane.

Twigs slender, zigzag; pith *finely chambered,* at least near nodes, the chambers hollow.

FIGURE 40   Winged elm, *Ulmus alata*. Enlarged twig with uppermost lateral bud, leaf
scar, and twig scar; leaf; twig section showing corky wings of bark.

Terminal bud absent; lateral buds small, pointed, *closely appressed to twig,* with several overlapping scales.

Leaf scars elliptical, half circular, or crescent shaped. Bundle scars three. Stipule scars present, very small.

Fruit a small, nearly globose drupe with a wrinkled stone, resembling a dry cherry, borne on twigs of current season; thin, sweet pulp edible; often persisting into winter.

1 Leaves mostly lanceolate, entire or with only a few teeth; buds less than
   3 mm long . . . . (1) *C. laevigata*
1' Leaves mostly broader than lanceolate, toothed, at least above middle; buds
   3 mm long or more. . . . . (2) *C. occidentalis*

| | 1. *Celtis laevigata* |
| FIGURE 41 | sugarberry |

Bark of trunk *light gray, smooth or warty.*

Leaves two-ranked, broadly lanceolate or narrowly ovate, 4–8 cm long, *usually more than twice as long as wide,* palmately veined, *entire* or sometimes with a few teeth, long pointed at apex, unsymmetrical at base, dark green above, paler beneath, glabrous or nearly so on both sides except for minute hairs along margin.

Twigs green turning reddish brown, mostly glabrous.

Buds *very small,* 1–3 mm long, ovoid, pointed, closely appressed, minutely hairy.

Fruits 6–8 mm across, orange red to yellow, the flesh sweet, hence name of tree.

Low, moist woods; Henderson and Madison Cos., NC; Oconee Co., SC; GSMNP.

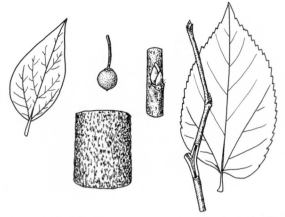

FIGURE 41   *Celtis.* Left: leaf of sugarberry, *C. laevigata.* Center and right: hackberry, *C. occidentalis.* Fruit, typical bark, lateral bud and leaf scar, leaf showing unequal base, and twig showing chambered pith.

2. *Celtis occidentalis*
FIGURE 41 | hackberry

Bark of trunk brownish gray, *roughened with knobs and narrow ridges.*

Leaves two-ranked, ovate, 5–15 cm long, palmately veined, *sharply toothed,* at least above middle, short pointed to very long pointed at apex, usually (but not always) unsymmetrical at base, dark green above, paler green beneath, smooth or rough on both sides.

Twigs grayish brown, usually glabrous; lenticels raised, elongated longitudinally, scattered.

Buds 2–4 mm long, ovate or triangular in outline, lying flat against twig, brown, hairy, with several overlapping scales.

Fruits about the size of a pea, orange red to dark purple, with thin flesh and large stone; often remaining on tree for several weeks after ripening; flesh sweet, edible, tasting like dates.

Rich woods; occasional, NC; rare, GSMNP.

Mulberry Family (Moraceae)

Leaves simple, usually alternate, in our species deciduous; roots frequently orange; sap milky; flowers small, staminate and pistillate on same plant or on separate plants, in catkinlike spikes or heads or in hollow receptacles; fruit multiple, consisting of all the ripened ovaries of each flower cluster; Osage-orange, paper-mulberry, mulberry.

	Maclura pomifera
FIGURE 42	Osage-orange

A low-branching, often crooked tree *armed with sharp thorns*. Bark of trunk orange brown, furrowed and shreddy (similar to that of *Robinia pseudoacacia*). Inner bark *bright yellow*.

Leaves alternate, often clustered on spur twigs, deciduous, simple, ovate, 7–15 cm long, entire, long pointed at apex, rounded at base, glossy green.

Twigs slender to moderately stout, zigzag, light green to olive gray, glabrous; short spur twigs common on older growth; thorns solitary, unbranched, needle sharp, *gradually decreasing in length toward twig tip*.

Terminal bud absent. Lateral buds small, often paired, flattened-globe shaped, *appearing partly sunken in bark, at the side of thorns*, with four or five brownish scales.

Leaf scars alternate, half circular to triangular, slightly raised. Bundle scars usually more than three. Stipule scars present, small.

Fruit a *rough, greenish ball* (multiple fruit), about the size of grapefruit, dry and pulpy, with numerous seeds (nutlets) and bitter milky juice, not edible.

Native of Arkansas, Oklahoma, and Texas, occasionally persisting after cultivation in our range.

	Broussonetia papyrifera
	paper-mulberry

A large shrub or small tree with smooth bark and *milky sap*.

Leaves sometimes opposite but typically alternate, two-ranked, deciduous, simple, broadly ovate, 8–20 cm long, *palmately veined*, coarsely toothed and often variously lobed, more or less unequal or unsymmetrical at base, grayish green and with *rough hairs* on upper surface, paler and softly hairy beneath.

Twigs moderate to stout, grayish green, hairy, *rough to touch*; pith white, with *a thin green partition* at each node.

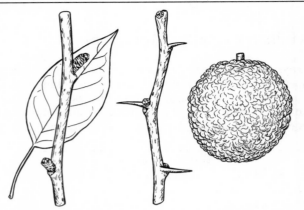

FIGURE 42   Osage-orange, *Maclura pomifera*. Leaf, older twig with characteristic short
growths, twig with thorns and a bud at base of each thorn, and fruit.

Buds conical, 4–5 mm long, with usually two scales visible, the outer scale, and
    sometimes the inner, marked with lengthwise lines (striate). Terminal bud
    absent.
Leaf scars mostly alternate, raised, elliptical to nearly circular. Bundle scars about
    five. Stipule scars present, long and narrow.
Pistillate plants (and therefore fruit) not seen in our area.
Native of China and Japan, occasionally escaped from cultivation, mostly in or near
    towns.

Morus (mulberry)

Small trees with palmately veined leaves and milky sap.

Leaves alternate, two-ranked, deciduous, simple, either lobed or unlobed,
coarsely toothed, broadly ovate to nearly circular in general outline, with *three or
more principal veins from base.*

Twigs slender to moderate, circular in cross section; *sap of twigs and petioles
milky* (perhaps not clearly evident in cut twig during winter unless temperature is
well above freezing).

Terminal bud absent. Lateral buds with mostly five to seven scales overlapping
in two ranks.

Leaf scars half circular to nearly circular, more or less *depressed in center.*
Bundle scars *numerous,* often raised and appearing as dotlike projections, arranged
in a closed ring or irregularly scattered. Stipule scars present, narrow.

Fruit fleshy, whitish to reddish or dark purple, resembling an elongated blackberry, edible; ripe in May or June.

1 Leaves smooth above, frequently lobed, the lobes rounded; lower surface of
   leaves hairy along veins but otherwise glabrous or with a few scattered hairs;
   buds 2–4 mm long, the scales uniformly brown or reddish brown
   . . . . (1) M. alba
1' Leaves rough above, seldom lobed; lobes, if present, pointed; lower surface
   of leaves usually velvety-hairy; buds mostly more than 4 mm long, the scales
   tinged with green . . . . (2) M. rubra

	1. *Morus alba*
FIGURE 43	white mulberry

A tree with *yellowish brown bark,* the trunks of old trees showing a distinctive
  yellowish color between the ridges.
Leaves ovate in general outline, 5–15 cm long, coarsely toothed and usually palmately lobed, *the lobes rounded or short pointed,* but often unlobed, palmately
  veined, pointed at apex, obliquely heart shaped, rounded, or straight across at
  base, *glabrous* and more or less *glossy* on upper surface, glabrous or with whitish hairs along veins beneath.
Twigs yellowish green to brownish gray, usually glabrous.
Buds triangular in outline, mostly less than 4 mm long, about as broad as long,
  somewhat flattened and appressed, blunt, with five or six *reddish brown* scales
  visible.
Fruits white, pink, or blackish, 1–2 cm long, sweet, edible.
Native of Asia, occasionally escaped from cultivation.

	2. *Morus rubra*
FIGURE 43	red mulberry

A small to medium-sized tree with *dark brown,* scaly-ridged bark.
Leaves broadly ovate to nearly circular, 7–15 cm long, sharply toothed, usually not
  lobed but sometimes palmately lobed on vigorous growth, *the lobes long
  pointed,* palmately veined, long pointed at apex, heart shaped to straight across
  at base, *dull green and rough to touch on upper side,* paler and densely to thinly
  hairy beneath.
Twigs green turning light brown or reddish brown, glabrous.
Buds ovoid, longer than broad, pointed, 3–6 mm long, somewhat divergent, *greenish brown* or brown, glabrous; visible bud scales mostly six or seven (four to
  eight), the inner scales greenish with a dark margin.
Fruits red, mostly 2–3 cm long, sweet, edible.
Unlobed leaves of this species are somewhat similar to those of *Tilia,* but the latter
  are not sandpapery on upper surface.
Rich woods at lower elevations; occasional, throughout.

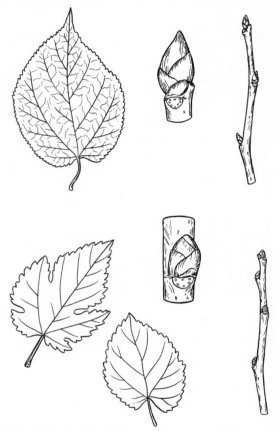

FIGURE 43   *Morus.* Above: red mulberry, *M. rubra.* Leaf, enlarged bud and leaf scar, twig. Below: white mulberry, *M. alba.* Two forms of leaves, enlarged lateral bud and leaf scar, and twig.

Sandalwood Family (Santalaceae)

Mostly deciduous shrubs, usually parasitic on roots of other plants; leaves simple, alternate or opposite, entire; flowers small, in terminal or axillary clusters or solitary; fruit a one-seeded nut or a drupe. One species is described for our area.

FIGURE 44	*Pyrularia pubera* buffalo nut, oilnut

A fairly large shrub.

Leaves alternate, deciduous, simple, thin, prominently veiny, oblong-ovate to broadly lanceolate, 5–15 cm long, entire, long pointed to rounded at apex, wedge shaped to rounded at base, softly hairy when young, glabrous at maturity.

Twigs moderate to rather stout, olive brown, hairy or glabrous; pith white, *spongy*. First-year twigs often self-pruned, leaving *large twig scars, each with many bundle scars* arranged in a closed ring.

Buds not evident on first-year twigs, *solitary or in irregular clusters* of two or more on older growth, widely divergent, ovoid, pointed, mostly 4–8 mm long, with several short scales at base and several longer greenish scales above; tips of upper scales usually reddish brown. Terminal bud lacking.

Leaf scars nearly circular to shield shaped. Bundle scar one, or sometimes three or more, sunken below surface of leaf scar. Stipule scars absent.

Fruit a *pear shaped* to nearly globose drupe, 20–25 mm long, with thin flesh and *large stone; poisonous.*

Moist woods at low and middle elevations; common, throughout.

Mistletoe Family (Loranthaceae)

Shrubs that are parasitic on branches and trunks of flowering trees; leaves opposite, persistent, entire; flowers small, in spikelike clusters; fruit a berry. Only one species occurs in our area.

Phoradendron flavescens American mistletoe

A many-branched shrub, *parasitic* on branches of oaks, black tupelo, and other deciduous trees.

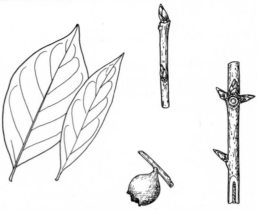

FIGURE 44   Buffalo nut, *Pyrularia pubera*. Two forms of leaves, twig showing solitary buds and leaf scars, fruit, and twig showing clustered buds, a branch scar, and spongy pith.

Leaves opposite, simple, evergreen, thick and leathery, palmately veined, oblong or obovate, 2–13 cm long, entire, rounded or blunt at apex, wedge shaped at base, yellowish green, glabrous or slightly hairy.

Twigs stout, greenish, brittle at base, often falling off and leaving twig scars.

Fruit a white berry, 4–6 mm across, containing two or three white seeds surrounded by sticky pulp.

Occasional, at lower elevations, throughout.

Birthwort Family (Aristolochiaceae)

Twining vines or herbs with simple, alternate leaves, mostly heart shaped, palmately veined, and entire; flowers small to large, in axils of leaves; fruit a capsule. In our area only one species is woody.

FIGURE 45	*Aristolochia durior* Dutchman's-pipe

A soft-wooded, twining vine with stems often long and high climbing.

Leaves alternate, deciduous, simple, *large,* palmately veined, very thin, *distinctly heart shaped to nearly circular,* 8–24 cm across, entire, green and glabrous above, finely hairy to nearly glabrous beneath.

FIGURE 45   Dutchman's-pipe, *Aristolochia durior.* Twining stem, leaf, and section of stem showing U-shaped leaf scar and superposed buds.

Stems green, glabrous, enlarged at nodes, without tendrils or aerial rootlets, dying back some distance from tip in autumn.

Buds small, covered with white hairs, *two or three together (superposed) on a silky pad,* enclosed by base of leaf petiole or surrounded on three sides by leaf scar; often with an additional bud above petiole base or leaf scar. Terminal bud absent.

Leaf scars narrowly U-shaped, with three bundle scars. Stipule scars absent.

Fruit a dry, cylindrical capsule, 5–8 cm long, containing many flat seeds.

Rich woods and banks of streams; fairly common, throughout.

Buttercup Family (Ranunculaceae)

Usually perennial herbs, only two genera in our area woody; leaves alternate or opposite, lobed, dissected, or compound; flowers regular or irregular, the sepals often colored and petallike; fruit a one-seeded follicle or cluster of achenes; yellowroot, clematis.

	Xanthorhiza simplicissima
FIGURE 46	yellowroot

A small, sparingly branched shrub, usually in colonies, spreading by underground stems. *Roots and inner bark of stems yellow,* with strong odor and bitter taste. Leaves and leaf scars crowded at top of short stem.

Leaves alternate, deciduous, once or (rarely) twice compound, long petioled, the *base of petiole extending partway around stem and clasping it.* Leaflets mostly five, lanceolate-ovate to broadly ovate, *irregularly toothed and often deeply cleft,* wedge shaped at base, bright green and more or less lustrous.

Twigs moderate in size, grayish, glabrous.

Terminal bud present, much larger than laterals, 10–18 mm long, slender pointed, slightly narrowed at base, with usually five (four to six) reddish green scales visible, the lower scales bearing awns. Lateral buds small and flattened, appressed to twig, with two or three scales showing.

Leaf scars narrow, shallowly U-shaped, *more than half encircling the stem.* Bundle scars about ten. Stipule scars lacking.

Fruit a small, dry one-seeded pod (follicle), in drooping clusters.

Shaded streambanks and damp woods at low elevations, throughout.

Clematis (clematis)

Soft-wooded vines, *climbing by means of twining petioles,* the latter *persistent* after leaf blades have withered and finally fallen.

Leaves opposite, pinnately compound; leaflets three to seven, toothed or lobed.

Stems soft and *scarcely woody,* dying back in winter; *six-angled* (in our species), without aerial rootlets. Pith white, angled or star shaped in cross section, solid but with a thin partition at each node.

Buds in axils of petioles, small, with two to six mostly hairy scales. Terminal bud lacking.

Leaf petioles opposite, usually persistent, concealing broadly V-shaped leaf scars, joined around stem by a transverse ridge. Bundle scars five to seven. Stipule scars absent.

Fruits small, seedlike achenes with long, feathery tails, clustered into conspicuous fluffy heads.

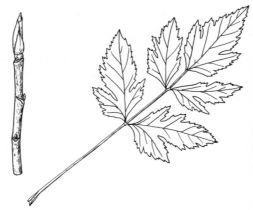

FIGURE 46   Yellowroot, *Xanthorhiza simplicissima*. Twig with large terminal bud, and compound leaf.

1 Leaflets mostly three, irregularly toothed or occasionally lobed; stems light brown or straw colored. . . . . (2) *C. virginiana*
1' Leaflets five to seven, some again divided, entire or lobed but not toothed; stems reddish brown . . . . (1) *C. viorna*

1. *Clematis viorna*
leather-flower

Leaflets five to seven, ovate to ovate-oblong, 3–8 cm long, two to three-lobed or entire but *not toothed,* or sometimes again divided, blunt or short pointed at apex, wedge shaped to somewhat rounded at base, bright green above, paler beneath, usually glabrous on both surfaces.
Stems climbing high over bushes and fences, nearly herbaceous, six-angled, *reddish brown,* thinly hairy or glabrous.
Moist woods and thickets; common, throughout.

2. *Clematis virginiana*
virgin's bower

Leaflets *three,* ovate, 2–10 cm long, coarsely toothed or lobed, long pointed at apex, rounded to somewhat heart shaped at base, glabrous on both sides or with a few scattered pale hairs beneath.
Stems sprawling or climbing, often nearly herbaceous, with six prominent ridges, making stems six-angled in cross section, *light brown or straw colored, hairy* or glabrous.
Woods and streambanks at lower elevations; common, throughout.

Barberry Family (Berberidaceae)

Herbs or small shrubs with alternate or clustered, deciduous, simple or compound leaves; flowers in clusters or solitary; fruit a berry or capsule. Represented in our area by one genus.

Berberis (barberry)

Shrubs *armed with simple or branched spines* formed by the modification of primary leaves.

Leaves alternate or in alternate clusters, deciduous, simple, toothed or entire, wedge shaped at base.

Twigs angled or ridged downward from nodes; *inner bark and wood bright yellow.*

Buds small, ovoid, enclosed by six to eight overlapping scales.

Leaves deciduous above base, leaving persistent remnant of petiole bearing half-circular leaf scar. Bundle scars usually three. Stipules and stipule scars absent.

Fruits ellipsoid, scarlet berries, often persistent through winter; sour but sometimes used for jelly.

1 Leaves entire; spines usually simple ....(1) *B. thunbergii*
1' Leaves remotely toothed; spines usually three-branched ....(2) *B. canadensis*

FIGURE 47 | **1. *Berberis thunbergii***
Japanese barberry

Leaves 1–3 cm long, *entire,* the apex blunt or rounded, the base tapering gradually to petiole, bright green above, paler beneath.

Twigs slender, strongly three-ridged downward from nodes, purplish red when young, purplish brown in second year, glabrous.

Spines mostly *simple,* 4–15 mm long, straight and slender, sharp; occasionally two to three branched at a node.

Berries 8–10 mm long.

Native of Japan, occasionally escaped from cultivation into open woods and along roadsides.

FIGURE 47 | **2. *Berberis canadensis***
American barberry

Leaves 1–6 cm long, *sharply toothed,* the teeth widely spaced, broadly rounded or blunt at apex, gradually tapering downward to a short petiole and often appear-

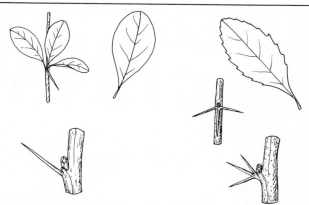

FIGURE 47   *Berberis.* Left: Japanese barberry, *B. thunbergii.* Leaves clustered on spur growth, a solitary leaf, and detailed section of twig showing bud and spine. Right: American barberry, *B. canadensis.* Leaf, and two detailed views of buds and branched spines.

ing without a petiole; upper surface green and glabrous, lower surface grayish green.

Twigs slender, angled, especially near tip, finely warty dotted, reddish to purplish brown aging to brown or gray; *spines usually three-pronged.*

Leaf scars small, half circular, much raised on top of persistent base of leaf petioles.

Berries 6–7 mm long.

Dry woods and bluffs; occasional, throughout.

Moonseed Family (Menispermaceae)

Climbing or twining vines, our species woody; leaves alternate, simple, palmately veined; flowers in clusters; fruit a drupe, the ovary often curving as fruit develops, causing seed to be bent into a crescent; coralbeads, moonseed.

	Cocculus carolinus coralbeads, Carolina snailseed

A *twining,* usually *sprawling* vine.

Leaves alternate, simple, *palmately veined,* 6–12 cm long, variable in shape, somewhat triangular to ovate, entire to three-lobed, tapering to pointed, rounded, or notched apex, heart shaped to straight across at base, dark yellowish green and slightly hairy above, paler and softly hairy beneath.

Stems slender, flexible, twining, grayish green to grayish brown, loosely hairy to glabrous.

Buds small, in slight depressions above leaf scars, concealed under dense pale wool.

Leaf scars half circular, often deeply depressed in center. Bundle scars three or more, indistinct. Stipule scars absent.

Fruits in clusters, *red,* glossy, globular drupes, 6–8 mm across, each containing a crescent-shaped seed.

Open woods and fields; Madison and Polk Cos., NC; Oconee and Pickens Cos., SC; GSMNP.

FIGURE 48	*Menispermum canadense* moonseed

A *twining* vine, climbing or sprawling.

Leaves alternate, deciduous, simple, *palmately veined,* nearly circular in outline, *large,* 6–20 cm long and wide, *variously three- to seven-lobed or -angled,* dark green above, paler and with a few hairs beneath; *petiole attached to leaf blade a short distance inside margin on lower side.*

Stems slender, flexible, rounded in cross section, *striate* with fine longitudinal ridges, greenish brown, hairy or glabrous.

Buds very small, hairy, partly buried in tissue of stem, several at each node, one (flower bud) in a hairy cavity above leaf scar, one or two other buds sunken beneath leaf scar.

Leaf scars large, elliptical or circular and split or notched at top, with raised margin and concave center. Bundle scars three to seven. Stipule scars none.

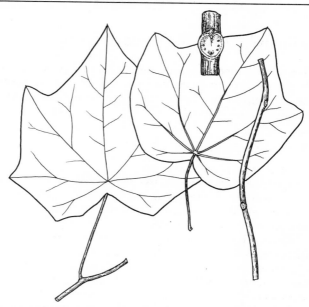

FIGURE 48   Moonseed, *Menispermum canadense.* Upper surface of leaf, lower surface of leaf showing attachment of petiole a short distance inside margin; section of stem showing leaf scar, sunken buds, and bundle scars; section of twining stem.

Fruits in drooping, grapelike clusters: globular, 8–15 mm across, dark blue with bloom, each with a single seed; flesh thick, juicy, not edible, *poisonous;* seeds *flattened, crescent shaped.*
Moist woods; occasional, throughout.

Magnolia Family (Magnoliaceae)

Trees with leaves alternate, simple, entire or lobed, mostly deciduous; bark aromatic; twigs encircled by stipule scars at nodes; flowers terminal, large, showy; fruit an aggregate of follicles or samaras, conelike; magnolia, yellow-poplar.

Magnolia (magnolia)

Trees with aromatic bark.

Leaves alternate, in our species deciduous, simple, *large, entire.*

Twigs *encircled at each node by a transverse or oblique scar left by fallen stipules;* fragrant when cut or broken but somewhat bitter to taste. Pith solid but sometimes with firmer partitions at intervals.

Terminal bud *large, enclosed in a single outer scale* (actually consisting of two united stipules). Lateral buds often much smaller.

Leaf scars nearly circular to broadly U-shaped, the surface showing *numerous,* usually *scattered bundle scars.*

Flowers large, showy, white to yellow or yellowish green; April–May.

Fruit a large, upright, fleshy "cone"; when mature, many red "seeds" (follicles) hanging on slender threads.

1 Leaves crowded near end of young twigs or, in the absence of leaves, the leaf scars similarly crowded on the somewhat swollen end of year's growth (may be scattered on vigorous twigs); terminal bud greenish to purplish, glabrous and often glaucous:

  2 Leaves deeply heart shaped or two-lobed at base, glabrous beneath
    ....(1) *M. fraseri*

  2' Leaves wedge shaped at base, finely hairy beneath, at least when young
    ....(2) *M. tripetala*

1' Leaves or leaf scars not crowded as above, but more or less uniformly spaced along young twigs; terminal bud whitish-silky..............(3) *M. acuminata*

FIGURE 49 | 1. *Magnolia fraseri*
              | Fraser magnolia

Bark smooth, light gray, becoming scaly on old trunks.

Leaves alternate but *crowded toward tip of first-year twigs,* obovate, 10–50 cm long, entire except for *two prominent lobes at base of blade,* rounded or pointed at apex, bright green above, paler beneath, glabrous on both sides.

Twigs moderately stout, reddish brown to greenish brown, glabrous. Stipule scars encircling twig at nodes.

FIGURE 49   Fraser magnolia, *Magnolia fraseri*. Leaf with two basal lobes; twig with
terminal bud and crowded leaf scars at end of each year's growth.

Terminal bud large, mostly 2 cm long or more, very slender and sharply pointed,
green to purple, glabrous and often glaucous. Lateral buds similar but smaller,
or absent.

Leaf scars crescent shaped to shield shaped, crowded on the somewhat swollen end
of year's growth, or sometimes more or less evenly distributed on vigorous
growth.

Rich woods; common, throughout.

	2. *Magnolia tripetala*
FIGURE 50	umbrella magnolia

Bark light gray and smooth.

Leaves *clustered near end of twigs,* suggesting common name, obovate to broadly
lanceolate, 10–45 cm long, entire or wavy margined, long pointed at apex, long
tapered at base, green and glabrous above, pale green and usually hairy beneath.

Twigs *stout,* reddish brown, glabrous or nearly so, lustrous; marked by stipular
rings at nodes; with fragrant odor if cut or broken; pith not chambered.

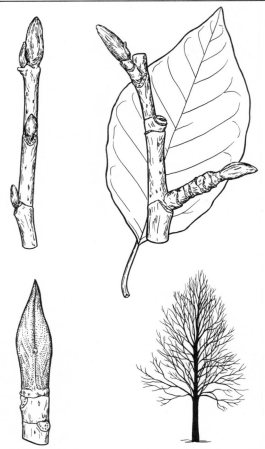

FIGURE 50   *Magnolia*. Above and lower right: cucumbertree, *M. acuminata*. Leaf; twigs
showing terminal buds, lateral buds, and leaf scars; and silhouette of mature
tree. Lower left: umbrella magnolia, *M. tripetala*. End of twig showing
terminal bud and stipule scars encircling twig at each leaf scar.

Terminal bud *very large,* 3–4 cm long, conical, the tip pointed and often curved,
    purplish, smooth and glaucous. Lateral buds very small or scarcely visible.
Leaf scars large, elliptical, mainly crowded at swollen places on twig.
Rich woods; occasional, NC, GA, GSMNP.

| FIGURE 50 | 3. *Magnolia acuminata*
cucumbertree |

Bark grayish brown; on old trees, separated into a network of furrows and scaly ridges.

Leaves *rather evenly spaced along twigs,* obovate, broadly elliptical, or ovate, 10–30 cm long, entire or somewhat wavy margined, short pointed or long pointed at apex, broadly wedge shaped or rounded at base, yellowish green and glabrous above, pale green and hairy or glabrous beneath.

Twigs moderately stout, light reddish brown to greenish brown, glabrous or slightly hairy, aromatic if cut or broken; stipule scars beginning at tips of leaf scars, appearing as transverse or oblique lines and completely encircling twigs at nodes.

Terminal bud long and slender, 1–2 cm long, *densely white-silky.*

Leaf scars crescent shaped to U-shaped, scattered along twigs (may be somewhat crowded on slow-growing twigs).

Rich woods; infrequent, throughout.

| FIGURE 51 | *Liriodendron tulipifera*
yellow-poplar, tuliptree |

A tall tree. Bark grayish, deeply and regularly furrowed on old trunks; on young trunks nearly smooth but very shallowly furrowed, and *whitened in the developing furrows.*

Leaves alternate, deciduous, simple, squarish in shape, 6–20 cm long and wide, mostly four-lobed (occasionally an additional pair of lobes toward base), *very broadly and shallowly notched at apex;* dark green and lustrous above, paler and glaucous beneath; turning clear yellow in autumn; stipules large and conspicuous.

Twigs moderately stout, reddish brown and glabrous when mature, commonly with purplish bloom; yielding a fragrant odor when cut or broken, and with bitter taste. Pith white and solid but chambered, showing firm green partitions at intervals.

Terminal bud present, *flattened and blunt,* mostly 10–15 mm long, with two valvate scales, green, purplish, or brownish red, glabrous. Lateral buds reduced in size, often two together (superposed), or absent.

Leaf scars large, elliptical to nearly circular. Bundle scars twelve or more, scattered. Stipule scars *completely encircling twig at nodes.*

Flowers large, cup shaped, tuliplike, with six green-and-orange petals, borne solitary and erect; April–June.

Fruit a narrow pointed "cone," 6–8 cm long, borne upright, made up of long-winged seeds (samaras); cone spreads open when ripe, then gradually breaks apart, leaving central stalk erect on stem.

Rich woods; common, throughout.

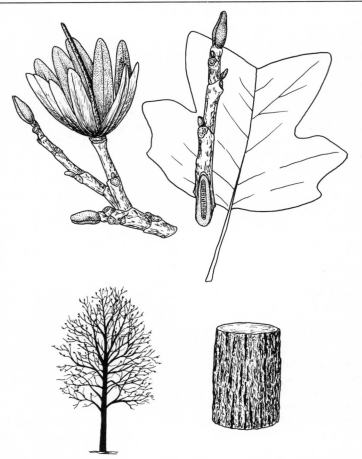

FIGURE 51   Yellow-poplar, *Liriodendron tulipifera*. Twigs and fruit cone after seeds (nutlets) have been shed; twig showing terminal bud, lateral buds and leaf scars, stipule scars encircling twig at nodes, and chambered pith; leaf; silhouette of mature tree; and section of trunk showing mature bark.

Calycanthus Family (Calycanthaceae)

Deciduous, aromatic shrubs with opposite, simple, entire leaves; flowers solitary, at ends of lateral stems; fruit a number of one-seeded achenes, enclosed within a large, fleshy, somewhat podlike receptacle. Only one genus is found in our area.

	Calycanthus fertilis
FIGURE 52	sweetshrub, smooth allspice

An upright shrub with *aromatic* leaves and twigs.

Leaves opposite, deciduous, simple, entire, lanceolate to ovate-lanceolate, 5–18 cm long, short pointed or long pointed at apex, rounded or broadly wedge shaped at base, more or less rough (scabrous) above, glabrous or with only scattered hairs beneath.

Twigs moderate in size, somewhat four-angled, *enlarged and flattened at nodes,* glabrous or nearly so.

Terminal bud absent. Lateral buds several in a single budlike group, small, *nearly or entirely concealed by base of leaf petioles* in summer, brown- to blackish-hairy, apparently without scales.

Leaf scars opposite, much raised, U-shaped, with three bundle scars. Opposite leaf scars not connected. Stipule scars none.

Flowers maroon or brownish, terminal, solitary, on short leafy branches; April–June.

Fruits large and leathery, baglike, 3–8 cm long, containing brown, seedlike achenes.

Low, moist woods and streambanks; common, throughout.

Calycanthus floridus (hairy sweetshrub) is similar to *C. fertilis* but has leaves densely hairy beneath and twigs with soft hairs persistent through winter. Along streams and on wooded slopes; occasional, NC, GA.

Custard-apple Family (Annonaceae)

Shrubs or small trees; leaves alternate, simple, entire, in our species deciduous; flowers solitary, large, purplish, usually in the axil of leaves of the previous season; fruit large, oval or oblong, smooth, fleshy, aromatic when mature.

FIGURE 52   Sweetshrub, *Calycanthus fertilis*. Leaves, fruit, and stem.

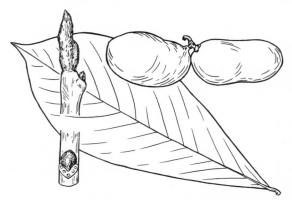

FIGURE 53   Pawpaw, *Asimina triloba*. Leaf, fruit, and enlarged sections of twig showing naked terminal bud, lateral bud, and typical leaf scar.

	Asimina triloba
FIGURE 53	pawpaw

A shrub or small upright tree.

Leaves alternate, deciduous, simple, 12–30 cm long, entire, broadest above the middle, gradually tapering toward base (resembling leaves of *Magnolia tripetala*), glabrous or nearly so on upper side, usually rusty-hairy beneath; crushed leaves with strong odor, like that of bell pepper.

Twigs moderately slender, slightly zigzag, reddish brown, sparingly rusty-hairy or glabrous; pith at first uniformly white, later chambered (*greenish partitions* alternating with sections of whitish or light brown tissue), and finally brown with chambers hollow.

Terminal bud present, *naked,* large and elongated, flattened and knifelike in outline (somewhat similar to that of witch-hazel, *Hamamelis virginiana*), brown-woolly or *reddish brown—woolly,* evidently without scales. Lateral buds densely woolly with reddish brown hairs and of two kinds: flower buds rounded, *stalked,* and leaf buds oblong, pointed, not stalked.

Leaf scars alternate, crescent shaped or horseshoe shaped, partly surrounding bud. Bundle scars usually five, rarely seven. Stipule scars none.

Fruit a kidney-shaped berry, 4–8 cm long, with yellowish green to brown skin and custardlike pulp surrounding several large flat seeds; pulp sweet and edible.

Low woods; occasional throughout.

Laurel Family (Lauraceae)

Aromatic trees or shrubs; leaves alternate, simple, entire or lobed, in our species deciduous; flowers small, yellow or yellowish green, usually fragrant; fruit a one-seeded drupe; spicebush, sassafras.

FIGURE 54 | *Lindera benzoin*
spicebush

An upright, *aromatic* shrub.

Leaves alternate, deciduous, simple, obovate, ovate, or oblong, 4–14 cm long, entire, short pointed or long pointed at apex, narrowly to broadly wedge shaped at base, glabrous or thinly hairy; with strong, pleasant, lemony odor when crushed; leaves decreasing in size toward base of twig, the lower leaves usually more rounded than the upper ones.

Twigs slender, *green to greenish brown,* glabrous.

Terminal bud absent. Lateral buds with about three visible scales and *of two kinds:* (*a*) leaf buds small, narrowly ovoid, appressed, often superposed, not stalked; and (*b*) flower buds nearly globose, *stalked, typically in pairs at a node,* one on each side of leaf bud, but often in clusters.

Leaf scars slightly raised, crescent shaped to half circular. Bundle scars three. Stipule scars lacking.

Fruit a *bright red,* ellipsoid drupe, 6–10 mm long, borne at nodes of last year's twigs, often persistent for some time after leaves have fallen.

Along streams and in rich moist woods; common, throughout.

FIGURE 55 | *Sassafras albidum*
sassafras

A large shrub or small tree; *all parts spicy-aromatic.* Mature bark *reddish brown,* deeply furrowed, with rough, broken ridges.

Leaves alternate, deciduous, simple, *variable* in size and form, 6–12 cm long (occasionally to 18 cm), *two-lobed or three-lobed, or sometimes unlobed,* the margin entire; mostly palmately veined, but some leaves pinnately veined; bright green to dull dark green above, paler and glabrous beneath (lower surface finely hairy in var. *molle*).

Twigs slender to moderate, brittle, usually some shade of *green,* sometimes reddish brown, glabrous (woolly-hairy in var. *molle*). Twigs *often branching in first year,* the main twig surpassed in length by the lateral twig just below it.

Terminal bud present, large, 6–9 mm long, green, often tinged with red or purple,

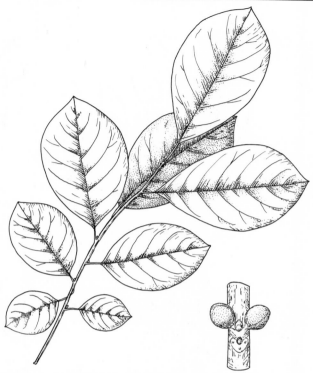

FIGURE 54   Spicebush, *Lindera benzoin*. Spray of foliage showing leaves decreasing in size toward base of twig; enlarged section of twig with stalked flower buds, leaf buds, and leaf scar.

glabrous or slightly hairy, with four to six scales visible. Lateral buds much smaller and with fewer scales.

Leaf scars alternate but irregularly spaced along twig, raised, half circular to nearly circular. Bundle scar *one,* appearing as a transverse line, or sometimes broken into three scars. Stipule scars absent.

Fruit an oblong, dark blue, fleshy drupe, 6–8 mm long, with a single bony stone, borne on a red, club-shaped stem; maturing in summer and soon falling.

Open woods, roadsides, old fields; common, throughout.

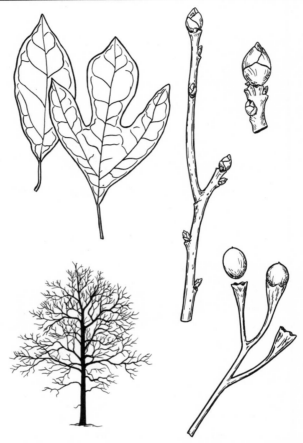

FIGURE 55   Sassafras, *Sassafras albidum*. Two forms of leaves, twig showing vigorous lateral growth in first year, enlarged terminal bud, silhouette of mature tree, and fruit and fruit stalks.

Saxifrage Family (Saxifragaceae)

Although largely herbaceous, this family is represented in our area by five woody genera; leaves alternate or opposite, deciduous, simple, lobed, toothed, or entire; flowers in few- to many-flowered clusters or racemes; fruit a capsule or berry with numerous seeds; mock-orange, hydrangea, gooseberry, and others.

	Itea virginica
FIGURE 56	sweet-spires, Virginia-willow

An erect shrub.

Leaves alternate, deciduous, simple, elliptical to obovate, 3–10 cm long, very finely toothed, short pointed or long pointed at apex, usually wedge shaped at base, green and glabrous above, paler and often sparingly hairy beneath.

Twigs mostly slender, circular in cross section, *green* or reddish green, hairy when young. Pith white, *chambered,* sometimes faintly, the chambers usually hollow.

Terminal bud present, conical, slightly larger than laterals. Lateral buds very small, less than 2 mm long, superposed, divergent, nearly globose, with three or four scales visible.

Leaf scars small, not or only slightly raised, half elliptical. Bundle scars three. Stipule scars none.

Fruit a narrow, two-part capsule, 3–7 mm long, containing many flattened seeds; borne in slender terminal clusters, persistent into winter.

Low woods; occasional, central mountains of NC, southward and westward.

	Decumaria barbara
	southern decumaria, climbing-hydrangea

A high-climbing vine with *aerial rootlets.*

Leaves opposite, deciduous, simple, ovate to ovate-oblong, 3–12 cm long, entire or sometimes coarsely toothed, short pointed or long pointed at apex, wedge shaped to heart shaped at base, glabrous and *shiny* above, glabrous or minutely hairy beneath.

Stems slender, circular in cross section, brown, glabrous or nearly so; older bark *loosening and peeling.*

Buds small, solitary, the scales indistinct, *reddish-hairy* (whereas buds of *Campsis radicans* are yellowish or greenish, glabrous, and scaly).

Leaf scars opposite, raised, *U-shaped;* paired leaf scars connected by a transverse line. Bundle scars three. Stipule scars lacking.

FIGURE 56   Sweet-spires, *Itea virginica*. Leaf, twig with chambered pith, and section of twig showing superposed buds.

Fruit a small capsule, strongly ribbed lengthwise, containing many seeds; in terminal clusters.

Low woods; Cherokee and Transylvania Cos., NC; Oconee and Pickens Cos., SC; GA.

Philadelphus (mock-orange)

Erect shrubs.

Leaves *opposite,* deciduous, simple, *three-veined from base,* toothed or entire.

Twigs slender, usually reddish brown, longitudinally lined or obscurely six-angled in cross section; *bark soon peeling off in thin, squarish bits and pieces.*

Buds solitary, small, with usually two scales. Terminal bud absent.

Leaf scars opposite, raised, crescent shaped to half circular. Bundle scars three. Stipule scars none, but each pair of leaf scars connected by a transverse line or ridge.

Fruit a more or less woody capsule, splitting into several sections when mature, containing numerous seeds.

1 Twigs glabrous; lateral buds concealed . . . . (1) *P. inodorus*
1′ Twigs hairy; lateral buds evident. . . . . (2) *P. hirsutus*

FIGURE 57 | 1. *Philadelphus inodorus*
               | scentless mock-orange

Leaves opposite, ovate to elliptical, 3–6 cm long, or sometimes to 10 cm on vigorous growth, entire or obscurely toothed, short pointed or long pointed at apex, rounded or broadly wedge shaped at base, glabrous and lustrous above, *glabrous or nearly so beneath,* except for hairs in axils of veins.

Twigs slender, glabrous, the bark checking and peeling in second year.

Buds *embedded under base of leaf petioles* or, in winter, *partly or entirely concealed* beneath membrane covering the leaf scars.

Leaf scars broadly crescent shaped to half circular, covered by a membrane, or membrane often ruptured by the underlying bud.

FIGURE 57   Scentless mock-orange. *Philadelphus inodorus.* Twig and leaves.

Capsule nearly globose, 6–8 mm across, usually in clusters of three; ripening in June
to August.
Streambanks and low hillsides; occasional, throughout.

2. *Philadelphus hirsutus*
hairy mock-orange

Leaves opposite, strongly three-veined from base, ovate to elliptical, 4–8 cm long,
finely toothed, long pointed at apex, mostly rounded at base, short-hairy or gla-
brous on upper surface, *softly hairy beneath.*
Twigs slender, reddish brown or straw colored, *hairy;* bark loose and peeling in
second year; flowering twigs short, usually with only one pair of leaves.
Buds *fully evident,* solitary, hairy.
Leaf scars broadly crescent shaped, not covering the buds.
Capsules obovoid, 5–7 mm across, solitary or in clusters of three (rarely four or
five), maturing in June to August.
Dry, wooded bluffs; occasional, throughout.

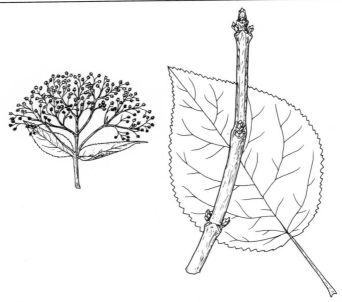

FIGURE 58   Wild hydrangea, *Hydrangea arborescens.* Umbrella-shaped cluster of fruits, leaf, and twig showing terminal bud, lateral buds, and leaf scars.

FIGURE 58	*Hydrangea arborescens* wild hydrangea

A shrub with *soft wood and large pith.* Bark *splitting and peeling* on older stems.

Leaves opposite, deciduous, simple, ovate, 4–18 cm long, rather coarsely toothed, short pointed or long pointed at apex, rounded or heart shaped at base, glabrous to densely hairy beneath; petioles mostly 2–11 cm long.

Twigs moderate in size, relatively stiff, brown, glabrous, glossy, six-sided in cross section, at least near tip of twig; bud scales persisting at base of young twigs.

Terminal bud present, ovoid, acute, 3–5 mm long, covered by two to four pairs of loose-fitting scales. Lateral buds with several pairs of scales, often short stalked.

Leaf scars opposite, broadly crescent shaped, each pair meeting around twig or connected by a transverse line. Bundle scars normally three, protruding. Stipule scars absent.

Flowers small, white, in flat-topped or umbrella-shaped terminal clusters, the outer flowers occasionally larger but sterile; May–July.

Fruit a small dry capsule, numerous in *flat- or round-topped terminal clusters,* often persistent in winter.

Moist, wooded cliffs and ravines, banks of streams; common, throughout.

Ribes (gooseberry, currant)

Unarmed or prickly shrubs with loose, shreddy outer bark.

Leaves deciduous, simple, *palmately veined and lobed,* the lobes toothed, alternate on young or fast-growing twigs, clustered at ends of short lateral spurs.

Twigs slender to moderate in size, prominently ridged downward from nodes, either with prickles, especially at base of leaf petioles (gooseberries) or without prickles (currants); bark soon cracking and peeling. Pith *spongy* in older twigs.

Buds ovoid or spindle shaped, stalkless or short stalked, with six or more loosely overlapping scales. Terminal bud present.

Leaf scars somewhat raised (in contrast to those of *Rosa,* which are flush with surface of twig), very narrow, U-shaped to crescent shaped. Bundle scars three. Stipule scars lacking.

Fruit a small, sour, juicy berry, sometimes prickly, containing many seeds; rarely present in winter.

1 Growth with strong unpleasant odor when crushed; stems reclining or
  sprawling, entirely unarmed. . . . . (1) *R. glandulosum*
1' Growth without strong unpleasant odor; stems erect or arching, armed with
  stiff prickles at some or all nodes:
  2 Berry prickly; nodal prickles 5 – 12 mm long . . . . (3) *R. cynosbati*
  2' Berry not prickly; nodal prickles usually short, 2 – 5 mm long, but some-
    times to 8 mm and often lacking. . . . . (2) *R. rotundifolium*

	1. *Ribes glandulosum*
FIGURE 59	skunk currant

A low shrub with *sprawling or reclining* stems, *not prickly.*

Leaves wider than long, mostly 3 – 8 cm wide, deeply five- to seven-lobed, coarsely
  double toothed, heart shaped at base, glabrous above, sparingly hairy beneath,
  with a heavy, unpleasant odor when crushed; hence common name.

Twigs slightly zigzag, ridged or lined downward from nodes, olive brown, glabrous
  or nearly so; prickles lacking; short growths numerous; inner bark with disagreeable odor.

Terminal bud cylindrical, sharp pointed, green to reddish purple, the scales loosely
  overlapping. Lateral buds appressed, *short stalked.*

Berry 5 – 8 mm across, dark red, glandular-bristly, scarcely edible.

Spruce-fir forests at higher elevations; NC, GSMNP.

FIGURE 59   Skunk currant, *Ribes glandulosum*. Leaves, fruit, and section of stem.

	2. *Ribes rotundifolium*
FIGURE 60	roundleaf gooseberry

An erect or spreading shrub.

Leaves alternate on leading twigs, clustered on short spurs, roundish in general outline, 2–5 cm across, three- to five-lobed; margin coarsely, irregularly, and
bluntly toothed; broadly wedge shaped, straight across, or heart shaped at base,
glabrous or finely hairy.

Twigs straw brown or yellowish brown to grayish brown, glabrous or minutely
hairy; *nodal prickles short,* usually less than 5 mm long, or prickles sometimes
lacking.

Buds slender, narrowly ovoid, pointed, 3–6 mm long, light brown to yellowish
brown, nearly glabrous, with six or seven scales exposed, the lower scales keeled.

Berry smooth, *not prickly,* 8–12 mm across, purplish.

Upland woods, open slopes, and balds at higher elevations, mostly above 4,500 feet;
NC, GA, GSMNP.

	3. *Ribes cynosbati*
FIGURE 60	pasture gooseberry, dogberry

An erect or spreading shrub.

Leaves alternate on long twigs, crowded on short lateral spurs, nearly circular in
outline, mostly 3–5 cm across, three- to five-lobed and irregularly toothed; tips

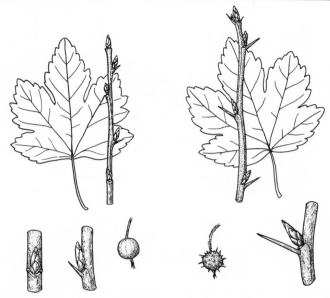

FIGURE 60   *Ribes.* Left: roundleaf gooseberry, *R. rotundifolium.* Leaf, twig, detailed sections of twig showing lateral buds and (right) nodal prickle; fruit. Right: pasture gooseberry, *R. cynosbati.* Leaf, twig, prickly fruit, and detailed section of twig showing lateral bud and nodal prickle.

of lobes blunt or pointed; straight across to heart shaped at base, softly hairy, at least when young.

Twigs yellowish to medium brown, glabrous, with slender prickles at nodes; *prickles mostly 5–12 mm long,* or sometimes absent; occasionally also with weak prickles between nodes.

Buds very slender, narrowly ovoid, pointed, mostly 5–7 mm long, golden brown to reddish brown, somewhat hairy to entirely glabrous, with six to eight visible scales, the lower scales keeled.

Berry *densely prickly,* 8–12 mm across, not including prickles.

Moist woods at middle and high elevations; NC, GA, GSMNP.

Witch-hazel Family (Hamamelidaceae)

Shrubs or small to large trees; leaves alternate, simple, deciduous, coarsely toothed or palmately lobed; flowers in small axillary groups, long terminal clusters, or dense heads; fruit a capsule, often woody; witch-hazel, sweetgum, fothergilla.

FIGURE 61	*Hamamelis virginiana* witch-hazel

A tall shrub or small tree. Bark thin, smooth or minutely scaly.

Leaves alternate, two-ranked, deciduous, simple, broadly elliptical to obovate, 5–15 cm long, coarsely wavy toothed, rounded, short pointed, or notched at apex, usually *unsymmetrical at base,* dark green above, paler beneath, covered at first with star-shaped clusters of hairs, later nearly glabrous or persistently hairy along lower veins.

Twigs slender, zigzag, light brown to orange brown, somewhat hairy or scurfy at first, becoming glabrous with age.

Terminal bud present, prominently *stalked and flattened,* brown-woolly, *without bud scales* but often partly covered by two scalelike stipules, these usually falling early. Lateral leaf buds similar but smaller. Flower buds small, globe shaped, in clusters of three, on short stalks, borne in axils of leaves.

Leaf scars two-ranked, small, half circular to three-lobed, the margins raised. Bundle scars usually three, occasionally more. Stipule scars present.

Flowers in small clusters, appearing late in autumn as last year's fruits ripen; each flower with four long, curled and twisted yellow petals, usually remaining, dried, well into winter.

Fruit a blunt woody pod (capsule), 10–15 mm long, containing two shiny black seeds; when ripe in autumn, forcibly expelling seeds for some distance; *empty gaping pods persisting in winter.*

Moist woods, slopes of ravines, banks of streams; common, throughout.

FIGURE 62	*Liquidambar styraciflua* sweetgum

A large tree. Mature bark rough, grayish brown, deeply furrowed, with narrow, rounded ridges.

Leaves alternate, deciduous, simple, 8–18 cm long and wide, palmately five- to seven-lobed and *star shaped,* the margin evenly toothed, bright green and lustrous above, pale green and glabrous beneath except for tufts of hairs in axils of veins; fragrant when crushed; petioles 6–15 mm long, slender.

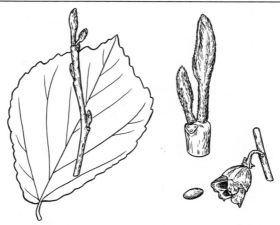

FIGURE 61　　Witch-hazel, *Hamamelis virginiana*. Leaf, twig, enlarged end of twig with stalked terminal bud, lateral bud, and leaf scar, and open capsule expelling seed.

Twigs moderately stout to slender, greenish to yellowish brown, glossy, often with *broad corky wings of bark* after the first year; short spur twigs common; sap resinous, aromatic. Pith *five-angled* in cross section.

Terminal bud present, large, 6–12 mm long, ovoid to conical, pointed, orange brown or reddish brown to green, shiny and resinous, six to eight scales fringed with hairs but otherwise glabrous. Lateral buds similar but smaller, divergent.

Leaf scars raised, crescent shaped to half circular. Bundle scars three, each appearing as a light-colored ring with a dark center. Stipule scars absent.

Fruit a long-stalked, *ball-shaped head* of beaked capsules, *woody and spiny,* containing small winged seeds; usually hanging on tree well into winter.

Moist or wet woods; occasional, throughout except at middle and high elevations.

FIGURE 63	*Fothergilla major* large fothergilla, witch-alder

An erect, few-branched shrub.

Leaves alternate, deciduous, simple, pinnately veined, *the lowest pair of lateral veins forming leaf margin for 2–10 mm on each side of midrib;* nearly circular to elliptical or obovate, 5–12 cm long, coarsely blunt toothed, rounded or short pointed at apex, heart shaped or straight across and usually symmetrical at base, glabrous or nearly so above, more or less hairy with star-shaped hairs beneath.

Twigs slender to moderately stout, zigzag, brown or reddish brown, dingy or scurfy with star-shaped clusters of hairs.

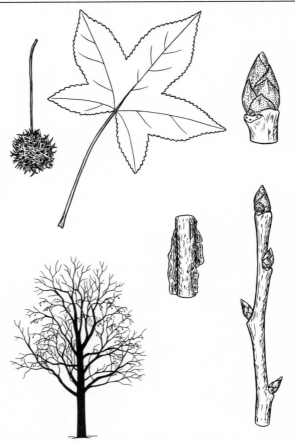

FIGURE 62   Sweetgum, *Liquidambar styraciflua*. Ball-shaped head of capsules, leaf, enlarged terminal bud, section of twig with corky wings of bark, twig with terminal bud, lateral buds, and leaf scars; silhouette of mature tree.

Buds *stalked,* ovoid or globose, brown, scurfy-hairy; outer bud scales two, falling early and leaving long narrow scars near base of bud; inner bud densely hairy.

Terminal bud present, ovoid, 5–11 mm long or more, including stalk. Lateral buds similar or somewhat smaller; one or two collateral buds sometimes present at a node. Flower buds with *a long narrow scale extending along one side of bud from base to tip.*

Leaf scars small, slightly raised, half circular or triangular with rounded corners. Bundle scars three. Stipule scars present, unequal in size and shape.

FIGURE 63   Large fothergilla, *Fothergilla major*. Leaf, terminal flower bud and uppermost leaf bud.

Flowers white, in dense terminal clusters, appearing in spring.
Fruit an ovoid capsule with a long beak, 10–12 mm long, including beak; in clusters, usually persistent in winter.
Dry woods and balds; Polk and Transylvania Cos., NC; Oconee Co., SC.

Plane-tree Family (Platanaceae)

Trees with deciduous bark and alternate, simple, palmately lobed leaves; petioles hollow at base, concealing buds; twigs encircled by stipule scars; flowers in heads; fruit a dense, globe-shaped head of achenes each with a tuft of hairs at base. Represented in our area by one genus.

FIGURE 64	*Platanus occidentalis* sycamore

A large tree with stout, widely spreading branches. Bark thin and mottled, falling away in brownish sheets of varying size and exposing the *greenish gray to creamy white underbark.* Old trunks often uniformly brown and scaly at base.

Leaves alternate, deciduous, simple, broad, 10–20 cm across, *palmately veined,* shallowly three- to five-lobed, the lobes triangular or occasionally not lobed, the margin coarsely toothed; lower surface coated with fuzz at first but soon becoming glabrous except on veins. Base of petiole *hollow, covering bud like a hood.* Stipules leaflike and prominent, the two together encircling twig.

Twigs moderately slender, zigzag, reddish brown to olive brown, *marked with a stipular ring at each node.*

Terminal bud absent. Lateral buds concealed under petioles until leaves are shed: conical, divergent, brown, *single scaled,* with a hint of folds in the scale.

Leaf scars collar shaped with scalloped margin, *nearly surrounding buds.* Bundle scars five to nine.

Fruit a *ball-shaped head,* borne singly on a slender stalk, consisting of many hairy achenes packed tightly together, persistent for a while but gradually breaking apart and scattering the "seeds."

Along streams, rich bottomlands, floodplains; infrequent, throughout.

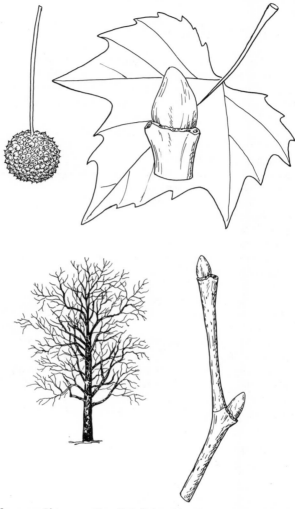

FIGURE 64   Sycamore, *Platanus occidentalis*. Ball-shaped head of fruits; leaf; enlarged topmost lateral bud, leaf scar, and twig scar; twig showing lateral buds, leaf scars, and stipule scar encircling twig at node; silhouette of mature tree.

Rose Family (Rosaceae)

A large family of woody and herbaceous plants; leaves alternate, usually with conspicuous stipules, simple or compound, deciduous or evergreen; flowers with mostly five sepals and petals; fruit various—a follicle, pome, hip, drupe, or an aggregate of small drupes; ninebark, spiraea, hawthorn, and many others.

FIGURE 65	*Physocarpus opulifolius* ninebark

A loosely branched shrub with *shreddy bark.*

Leaves alternate, deciduous, simple, roundish or ovate in outline, 3–7 cm long, 2–7 cm wide, larger on vigorous growth, more or less *three-lobed* or a few leaves unlobed, irregularly toothed, with three principal veins from base, dark green above, paler beneath, glabrous on both sides.

Twigs moderately slender, zigzag, orange brown to dark yellowish brown, strongly ridged downward from nodes. Older bark *peeling in thin strips,* hence common name.

Buds appressed, pointed, often twisted, 4–7 mm long, with about five overlapping scales visible. Terminal bud present, unless replaced by flower cluster or fruit cluster.

Leaf scars raised, three-sided or three-lobed, often somewhat shriveled or with ragged edges. Bundle scars three to *five, the upper pair very small, the middle one largest.* Stipule scars present, small, on each side at top edge of leaf scar.

Fruit a small dry, bladdery pod (follicle), borne in umbrella-shaped terminal clusters, usually persistent for a while.

Bluffs, moist slopes, ravines, streambanks; occasional, throughout.

Spiraea (spiraea)

Small shrubs with *slender,* erect stems.

Leaves alternate, deciduous, simple, toothed.

Twigs slender, more or less zigzag, circular in cross section or often ridged near nodes.

Buds small, single or collateral, with six or more scales exposed. Terminal bud usually absent, twig tips dying back to topmost lateral bud, but some twigs bearing terminal flower clusters or fruit clusters.

Leaf scars *very small,* half circular to crescent shaped, *usually much raised,* often indistinct or appearing partially torn. Bundle scar *one.* Stipules and their scars absent.

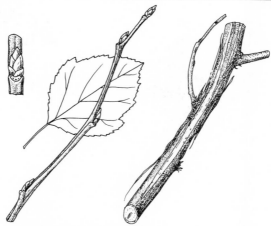

FIGURE 65   Ninebark, *Physocarpus opulifolius*. Lateral bud and leaf scar; leaf; twig; and
section of older stem showing bark peeling in thin strips.

Fruits small dry, podlike follicles, usually in groups of five, borne in showy terminal clusters, commonly persistent through winter.

1 Young twigs coated with rusty wool; lower surface of leaves woolly with
  whitish or brownish hairs. . . . . (1) *S. tomentosa*
1′ Young twigs not coated with rusty wool but somewhat hairy or without
  hairs; lower surface of leaves not woolly:
  2 Leaves entire, at least below middle, or often with only a few teeth near tip
    . . . . (2) *S. virginiana*
  2′ Leaves toothed from near base:
    3 Leaves long pointed at apex; flowers and fruits in roundish- or flattish-
      topped clusters. . . . .(3) *S. japonica*
    3′ Leaves short pointed or blunt at apex; flowers and fruits in pyramid-
      shaped clusters:
      4 Twigs yellowish brown or light brown; leaves finely toothed; stems of
        flower clusters and fruit clusters minutely hairy. . . . .(4) *S. alba*
      4′ Twigs reddish brown to purplish brown; leaves coarsely toothed;
        stems of flower clusters and fruit clusters glabrous . . . . (5) *S. latifolia*

FIGURE 66 | **1. *Spiraea tomentosa***
            steeplebush spiraea, hardhack

Leaves oblong to ovate or lanceolate, 2–7 cm long, 1–3 cm wide, single or double
toothed, short pointed or blunt at apex, wedge shaped to rounded at base, green
and glabrous above, *densely woolly beneath.*

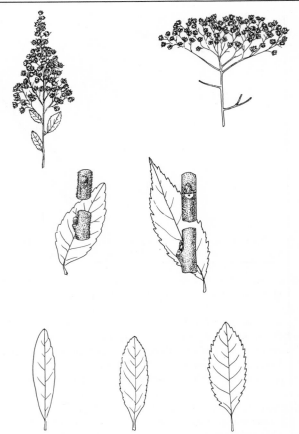

FIGURE 66   *Spiraea.* Upper left; steeplebush spiraea, *S. tomentosa.* Pyramid-shaped fruit
cluster, leaf, and detailed sections of twig showing buds and leaf scars.
Upper right: Japanese spiraea, *S. japonica.* Flat-topped cluster of fruits, leaf,
and detailed sections of twig showing lateral buds and leaf scars. Below, left
to right: leaves of Virginia spiraea, *S. virginiana,* narrowleaf spiraea, *S. alba,*
and broadleaf spiraea, *S. latifolia.*

Twigs sparingly branched, nearly straight, *woolly with rusty or grayish hairs,*
bronze brown beneath the wool; bark often loosening and peeling.
Buds very small, 1–2 mm long, ovoid, usually so woolly that the several scales are
indistinct.
Leaf scars very small, much raised, half circular, often appearing partially torn.
Flowers and fruits in long, rather *narrow clusters.*
Wet meadows, edges of woods; occasional, throughout.

| | 2. *Spiraea virginiana* |
| FIGURE 66 | Virginia spiraea |

Leaves *narrower* than those of other Spiraeas, oblong-lanceolate or oblanceolate, 3–6 cm long, mostly less than 15 mm wide, *entire or with only a few shallow teeth* above middle, short pointed to blunt with a small mucro at apex, wedge shaped at base, glabrous, glaucous beneath.

Twigs more or less angled, hairy when young, glabrous or nearly so at maturity, often glaucous.

Flowers and fruits in *flat-topped or rounded clusters.*

Streambanks; Ashe, Buncombe, Graham, Macon Cos., NC.

| | 3. *Spiraea japonica* |
| FIGURE 66 | Japanese spiraea |

Leaves lanceolate to ovate-lanceolate, 5–12 cm long, sharply toothed from near base and often double toothed, *tapering to a long point at apex,* wedge shaped at base, glabrous or nearly so, often *glaucous on lower side.*

Twigs brown to reddish brown, *covered with pale hairs,* at least near tip.

Buds very small, 1–2 mm long, rounded or triangular in outline but somewhat flattened.

Leaf scars very small, prominently raised above twig surface, with a pale border.

Flowers and fruits in flat-topped or slightly rounded clusters.

Native of Japan, established as an escape from cultivation; occasional, throughout.

| | 4. *Spiraea alba* |
| FIGURE 66 | narrowleaf spiraea, meadowsweet |

Leaves oblong to oblanceolate, 3–6 cm long, mostly three to four times longer than wide, *finely toothed,* bluntly pointed or rounded at apex, wedge shaped at base, green and essentially glabrous on both sides.

Twigs marked with lengthwise lines and angled near nodes, *yellowish to brownish,* nearly glabrous.

Buds ovoid, about 1 mm long, sparingly hairy near tip.

Flowers and fruits in *long, slender clusters;* stems of clusters more or less *hairy.*

Wet meadows and streambanks; occasional, NC.

| | 5. *Spiraea latifolia* |
| FIGURE 66 | broadleaf spiraea, meadowsweet |

Leaves generally elliptical, 2–7 cm long, two to three times longer than wide, *coarsely toothed,* short pointed to blunt at apex, wedge shaped or rounded at base, green and nearly glabrous on both sides.

Twigs marked with fine parallel lines and angled near nodes, *reddish to purplish,* glabrous or mostly so.

Buds globose to ovoid, about 1 mm long, reddish brown.

Flowers and fruits in *long, slender clusters;* stems of clusters *glabrous.*

Moist or dry uplands, streambanks; occasional, NC.

Aronia (chokeberry)

Erect shrubs.

Leaves alternate, deciduous, simple, toothed, with a row of *small, dark glands along midrib on upper side* (best seen under magnifying lens).

Twigs moderate to slender, rounded.

Terminal bud *pinkish to dark red, prominently flattened,* sharply pointed; visible bud scales four or five, somewhat keeled, the lower scales tending to be notched at apex, with a single dark spiny tip flanked by two teeth, or the spiny tip replaced by scar; upper bud scales pointed or spiny tipped. Lateral buds flattened, more or less appressed to twig, often reduced in size or lacking.

Leaf scars long and narrow, broadly U-shaped, constricted between three bundle scars. Stipule scars none.

Fruit a small, berrylike pome, in rounded or flattened clusters; edible but puckery, hence name chokeberry; persisting into winter.

1 Young twigs woolly; lower surface of leaves densely hairy to nearly glabrous
.... (1) *A. arbutifolia*
1' Young twigs and lower surface of leaves glabrous......... (2) *A. melanocarpa*

FIGURE 67	1. *Aronia arbutifolia* red chokeberry

Leaves elliptical to oblong-lanceolate or obovate, 4–10 cm long, finely toothed, the teeth either blunt or sharp, often incurved; short pointed to abruptly long pointed at apex, mostly wedge shaped at base; dark green to yellowish green and glabrous above, except for *reddish or blackish glands along upper midrib;* lower surface densely woolly to almost glabrous.

Twigs slender, rounded, *grayish-woolly,* pinkish or reddish beneath the wool.

Terminal bud 4–8 mm long, sharp pointed, distinctly *flattened,* usually more or less hairy but occasionally glabrous; bud scales four or five, keeled, pinkish red to deep red, the lower scales three-toothed, the middle tooth an extension of the keel but often replaced by scar.

Fruits *red,* 5–9 mm across.

Low woods, thickets, occasionally upland woods; infrequent, throughout.

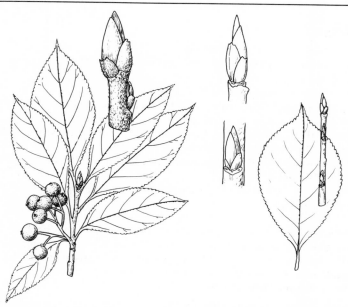

FIGURE 67   *Aronia*. Left: red chokeberry, *A. arbutifolia*. Twig with leaves, fruit cluster, and terminal bud; enlarged end of twig showing terminal bud and uppermost lateral bud. Right: black chokeberry, *A. melanocarpa*. Leaf, twig, enlarged terminal bud, enlarged lateral bud.

| | 2. *Aronia melanocarpa* |
| FIGURE 67 | black chokeberry |

Leaves elliptical to obovate, 3–9 cm long, finely toothed, abruptly long pointed to blunt at apex, broadly wedge shaped to rounded at base, bearing small, reddish brown glands along midrib on upper side, but otherwise glabrous.

Twigs slender, rounded to somewhat flattened, red or brownish red, *glabrous* or with a few scattered hairs; often with silvery coating.

Terminal bud 5–9 mm long, strongly *flattened,* sharply pointed, dark red or purplish red, glabrous; bud scales about four, keeled, the lower scales double notched and irregularly toothed. Lateral buds usually smaller.

Fruit dark purple to *black,* 7–10 mm across.

Wet woods, open woods, low heath balds; infrequent, throughout.

Crataegus (hawthorn)

Small trees or shrubs, mostly *thorny*.

Leaves alternate, deciduous, simple, pinnately or palmately veined, toothed and often lobed.

Twigs usually slender, more or less zigzag.

Thorns (modified twigs) borne in the axils of leaves as well as on older growth; *stiff and strong, sharply pointed,* usually simple, rarely branched.

Buds small, *rounded or globe shaped,* covered by four to eight shiny, reddish or brownish scales. Terminal bud present.

Leaf scars somewhat raised, narrow, crescent shaped. Bundle scars three. Stipule scars present, small.

Fruit a small, *applelike pome or haw,* containing two to five *bony* seeds (nutlets); often persisting for a while.

The hawthorns are fairly easy to recognize as a genus, especially if the thorns are present. However, most plants in our area are nearly impossible for anyone not a specialist in the group to identify. Flower and fruit characters are needed for identification, and I have not attempted to construct a key to the species.

Two of the common species are described below.

FIGURE 68	*Crataegus crus-galli* cockspur hawthorn

A shrub or small tree.

Leaves 3–6 cm long, widest and most clearly *toothed above the middle;* petioles 0–5 mm long.

Twigs slender, with many thorns 5–9 cm long.

Fruits 10–14 mm across, red or greenish.

Thickets, pastures, streambanks; common, throughout.

FIGURE 68	*Crataegus punctata* large-fruited hawthorn

A small tree or shrub.

Leaves 3–10 cm long, sharply double toothed, often shallowly *lobed* above the middle; petioles 10–25 mm long.

Twigs moderate to stout, with slender thorns 2–7 cm long, or sometimes nearly thornless.

Fruits 15–20 mm across, yellow or brown, *marked with brown dots.*

Stream bottoms and rich woods; common, throughout.

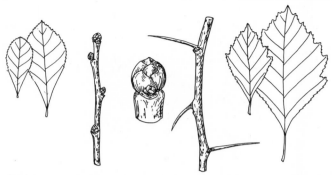

FIGURE 68   *Crataegus*. Left: leaves of cockspur hawthorn, *C. crus-galli*. Right: leaves of
              dotted hawthorn, *C. punctata*. Center, left to right: thornless twig, enlarged
              terminal bud, thorny twig, all representative of *Crataegus* species.

	Sorbus americana
FIGURE 69	American mountain-ash

A small upright tree or large shrub with smooth bark.

Leaves alternate, deciduous, *pinnately compound.* Leaflets thirteen to seventeen,
lanceolate, single or double toothed, short pointed or long pointed at apex,
wedge shaped or rounded at base, glabrous or nearly so, dark green on upper
side, paler green beneath, with or without a short petiole.

Twigs moderate to stout, brownish red to yellowish brown, or sometimes green
early in season, often covered with grayish skin; few branched, but spur twigs
common; with mild cherrylike odor when broken.

Terminal bud present, *large,* 9–18 mm long, oblong or conical, pointed, *red to
purplish red,* glabrous except at tip and along margin of scales, more or less
gummy, with three or four scales exposed. Lateral buds smaller, tightly appressed to twig, or absent.

Leaf scars crescent shaped to broadly U-shaped, *raised on persistent, dark red base
of petiole,* this later deciduous, exposing the true leaf scar. Bundle scars five,
arranged in a single curved line. Stipule scars none.

Fruit a *small orange red,* berrylike pome, 7–8 mm across, in heavy clusters, maturing in early autumn and persisting into winter.

Balds and high mountain forests, mostly above 5,000 feet; NC, GA, GSMNP.

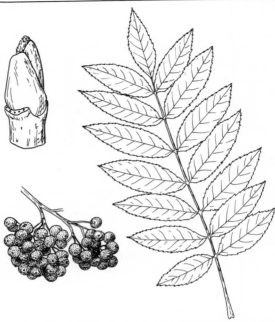

FIGURE 69   American mountain-ash, *Sorbus americana*. Enlarged terminal bud and leaf
scar, clusters of fruits, and pinnately compound leaf.

Amelanchier (serviceberry)

Shrubs or small trees with smooth grayish bark.

Leaves alternate, deciduous, simple, pinnately veined, sharply toothed.

Twigs slender, with odor of bitter almonds when bark is broken, but odor not
so strong as in the cherries (*Prunus*), and with bitter taste.

Terminal bud *long, slender, and sharply pointed,* somewhat similar to that of
beech (*Fagus grandifolia*) but with fewer scales visible; bud scales about six, more
or less keeled, often fringed with hairs along edge, *delicately colored, pink, reddish,
or greenish,* the lower scales three-toothed, the middle tooth often extended into a
spiny tip or sometimes replaced by a scar; upper scales sharp tipped. Lateral buds
similar on rapidly grown twigs, *curved toward twig,* much smaller or absent on
slowly grown twigs.

Leaf scars raised, crescent shaped or U-shaped, constricted between three bundle
scars. Stipule scars absent.

Fruit a small, berrylike pome, sweet, edible; remnants of fruits, dried and
shrunken, may be retained into winter.

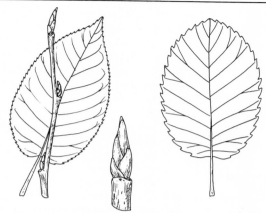

FIGURE 70   *Amelanchier.* Left: downy serviceberry, *A. arborea.* Leaf, twig, and enlarged terminal bud. Right: leaf of roundleaf serviceberry, *A. sanguinea.*

1 Leaves with six to ten teeth per centimeter of margin; lateral veins curving forward and branching, some of the smaller branches reaching the teeth; common species .... (1) *A. arborea*
1' Leaves with four to six teeth per centimeter of margin; lateral veins straight or nearly so, simple or once branched, extending into the teeth; rare species .... (2) *A. sanguinea*

	1. *Amelanchier arborea*
FIGURE 70	downy serviceberry

A tall shrub or tree where conditions are favorable, but commonly only a low shrub in dry woods.

Leaves elliptical to ovate or ovate-oblong, mostly 5–10 cm long, *sharply and finely toothed,* short pointed to *long pointed at apex,* more or less *heart shaped at base,* dark green and glabrous above, lower surface densely white-hairy to nearly glabrous when young, becoming less hairy or glabrous at maturity.

Twigs slender, slightly to moderately zigzag, nearly circular in cross section, reddish brown, glabrous.

Terminal bud 6–14 mm long, greenish to reddish brown, often tinged with pink, with five or six scales visible. Lateral buds similar, recurved toward twig.

Flowers white or pale pink, in terminal clusters on slender stalks; March–May.

Fruits 6–10 mm across, red or purplish red.

Dry or moist woods; common, throughout.

FIGURE 70 2. *Amelanchier sanguinea*
roundleaf serviceberry

A slender shrub with one to several stems from base.

Leaves widely oblong to elliptical, 3–7 cm long, *coarsely toothed,* rounded to blunt at apex, rounded to somewhat heart shaped at base, dark green above, paler beneath; lateral veins prominent.

Buds 6–10 mm long, very slender, long pointed, pink or red to greenish.

Fruits globose, 6–8 mm across, dark purple at maturity, with bloom.

Open woods, rocky slopes; Buncombe Co., NC.

Malus (apple)

Shrubs or small trees, commonly with *many short spur twigs,* the latter usually bearing crowded leaves or leaf scars and sometimes having a thorny tip.

Leaves alternate, deciduous, simple, toothed and sometimes lobed.

Twigs slender to rather stout, with little or no detectable odor when broken, but bark mildly bitter.

Terminal bud present, larger than the laterals, with four or five scales exposed. Lateral buds more or less flattened and appressed to twig.

Leaf scars raised, narrow, crescent shaped to broadly U- or V-shaped. Bundle scars three, often indistinct. Stipule scars very small or absent.

Fruit a small to large *apple.*

1 Leaves thinly hairy or nearly without hairs when mature; plants often thorny, the thorns consisting of sharply pointed spur twigs; buds conical-oblong:

  2 Leaves of short spurs with length usually less than twice their width; base of leaves mostly rounded to heart shaped; terminal bud 3–5 mm long
. . . . (1) M. *coronaria*

  2′ Leaves of short spurs with length usually more than twice their width; base of leaves mostly wedge shaped; terminal bud 1–3 mm long
. . . . (2) M. *angustifolia*

1′ Leaves permanently woolly beneath; plants never thorny; terminal bud 3–8 mm long, ovoid, blunt . . . .(3) M. *sylvestris*

FIGURE 71 1. *Malus coronaria*
sweet crab apple

A small tree, often forming thickets.

Leaves commonly *somewhat maplelike* but alternate, ovate-lanceolate or ovate to triangular-ovate, 4–9 cm long, either pinnately or palmately veined, toothed or

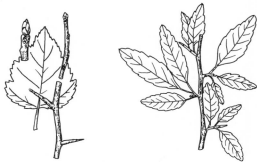

FIGURE 71   *Malus.* Left: sweet crab apple, *M. coronaria.* Leaf, terminal bud, and parts of twig, the lower one with thorn-tipped lateral twigs. Right: southern crab apple, *M. angustifolia.* Twig with leaves clustered on thorn-tipped lateral growths.

often with *a few small, triangular lobes near widest part,* short pointed or long pointed at apex, rounded or heart shaped at base, dark green and somewhat glossy above, paler and nearly glabrous beneath.

Twigs slender to moderate, reddish brown, covered in places with grayish skin, glabrous or sometimes hairy but soon becoming glabrous; *sharp-pointed lateral twigs common,* usually bearing leaves or leaf scars.

Terminal bud about 5 mm long, considerably larger than laterals, narrowly ovoid or conical, pointed, reddish brown to red, with four or five scales visible; bud scales fringed with hairs but otherwise glabrous, more or less keeled, often two-toothed, the tip of the keel extending between the teeth or keel sometimes replaced by a scar. Lateral buds smaller, about 3 mm long, somewhat flattened.

Fruit yellowish green or green, smooth and waxy to touch, about 3 cm across, fragrant, very acid, borne on a long, slender stalk.

Roadsides, fencerows, cleared areas, edges of woods; occasional, NC, GA, GSMNP.

	2. *Malus angustifolia*
FIGURE 71	southern crab apple, narrowleaf crab apple

A small tree with stiff, spreading branches, usually armed with sharply pointed spur twigs.

Leaves typically lanceolate, oblong, or narrowly elliptical, *mostly more than two times as long as wide,* 3–8 cm long, distinctly to obscurely toothed or nearly entire (leaves of vigorous growth often with several lobes or lobelike teeth), *rounded or blunt at apex,* wedge shaped or slightly rounded at base, dull green above, pale green beneath, woolly beneath when young, later glabrous or some hairs persisting.

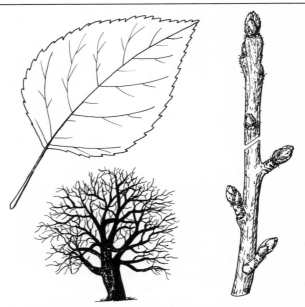

FIGURE 72   Apple, *Malus sylvestris*. Leaf, silhouette of mature tree, and detailed sections
of twig with terminal bud, lateral buds, and several fruiting spurs, each
ending in a terminal bud.

Twigs slender to stout, reddish brown, densely hairy when young but essentially
glabrous at maturity, or rapidly grown twigs sometimes remaining hairy; *spur
twigs often sharp and thornlike.*

Terminal bud narrow, blunt, 1–3 mm long, chestnut brown to brownish red, glabrous or more or less hairy, with four scales visible. Lateral buds similar but
smaller.

Fruit small, green or yellowish green, 25–35 mm across, with hard, sour flesh.

Fencerows, old fields, edges of low woods; occasional, throughout.

	3. *Malus sylvestris*
FIGURE 72	apple

A small, widely spreading tree.

Leaves broadly ovate to elliptical-ovate, 5–9 cm long, sharply or bluntly toothed
but *not lobed,* abruptly short pointed at apex, rounded or slightly heart shaped
at base, dark green and glabrous or nearly so above, *permanently woolly on
lower surface.*

Twigs moderately stout, dark reddish brown, *gray-woolly,* at least near tip, with

FIGURE 73   Pear, *Pyrus communis*. Leaf, sections of twig, and enlarged terminal bud.

fine lines or ridges descending from nodes; short, thick fruiting spurs common on older growth.

Terminal bud 3–8 mm long, much larger than the lateral buds, ovoid, blunt, reddish beneath *a light to heavy coating of gray hairs;* visible bud scales four or five but often obscured by hairs. Lateral buds small, more or less woolly, flattened against twig, with about three scales exposed.

Fruit variable in size, shape, and color; borne on a short, thick stalk.

Native of Europe and western Asia, occasionally escaped or persistent after cultivation.

	Pyrus communis
FIGURE 73	pear

A small tree with nearly erect branches and a narrow crown.

Leaves alternate but often crowded on short growths, deciduous, simple, broadly ovate to elliptical, mostly 4–8 cm long, finely and evenly toothed with blunt, rounded, or incurved teeth, short pointed or long pointed at apex, rounded at base, dark green and lustrous above, paler beneath, *glabrous on both sides* when mature; petioles sometimes as long as blade or longer.

Twigs moderate to stout, dark brown to yellowish green, glabrous; *stout spur twigs usually present;* some may be *sharp pointed and thornlike.*

Terminal bud present, 3–6 mm long, *conical, pointed,* reddish brown, *glabrous,* with four to six overlapping scales; bud scales more or less keeled, ending in a short, hard point, or the lower scales two-notched, the tip of the keel extending between two teeth (under magnifying lens). Lateral buds similar but smaller.

Leaf scars raised, crescent shaped. Bundle scars three. Stipule scars lacking.

Fruit a greenish pear, small on wild trees, seldom more than 6 cm long; flesh dry and gritty in wild form; ripe in August to October.

Native of Eurasia, rarely escaped or persistent after cultivation.

Rubus (blackberry, dewberry, raspberry)

Woody plants with erect, arching, or trailing stems, usually armed with prickles or bristles or both.

Stems or canes mostly *biennial,* in first year unbranched and not flowering (primocanes), in second year developing short lateral stems, bearing flowers and fruit, and then dying at end of season (floricanes), but the roots persisting and sending up new canes each year.

Leaves alternate, deciduous or half evergreen, simple or with three to five leaflets; stipules attached to petiole, leaving no scar on stem but often persisting on petiole remnant.

Buds commonly two together (superposed), the lower one smaller and covered by base of petiole, each bud with overlapping scales. Terminal bud lacking.

Leaves *deciduous above base of petiole,* leaving the basal part persistent on stem, this part appearing as a shriveled stub. Bundle scars three.

Fruit a "berry" (not a true berry, but an aggregate of small drupes).

The genus is made up of two distinct groups:
Ripe fruit, when picked, leaves a cone-shaped receptacle on plant: raspberries.
Ripe fruit, when picked, includes pulpy receptacle: dewberries and blackberries.
     Old stems trailing: dewberries.
     Old stems erect or arching: blackberries.

This is a large and complex genus. Some species, especially among the blackberries, are distinguished with great difficulty. Because crossbreeding is extensive, some of the plants in our area are almost impossible to name. You may come upon individual plants that do not fit any description closely.

1 Leaves simple, palmately three- to five-lobed; bark shreddy or peeling in long
   strips (flowering raspberry) .... (1) *R. odoratus*
1' Leaves compound, consisting of three or more leaflets:
   2 Leaves whitish-woolly beneath (raspberries):
     3 Stems armed with short stiff prickles, glabrous, whitened with bloom,
       purplish where bloom is removed .... (2) *R. occidentalis*
     3' Stems bristly with stiff glandular hairs:
       4 Bristly hairs mostly 1–2 mm long; stems upright or rarely arching,
         mostly without prickles, ashy-woolly beneath the bristly hairs, especially when young....................... (3) *R. idaeus* var. *canadensis*

4′ Bristly hairs mostly 3–5 mm long; stems arching, often rooting at tips, armed with reddish prickles, glabrous or only thinly hairy beneath the bristly hairs. . . . .(4) *R. phoenicolasius*

2′ Leaves green beneath:

  5 Stems prostrate or trailing, with only flowering and fruiting stems erect (dewberries):

    6 Leaflets thick, glossy dark green on upper surface; prickles slender, bristlelike, not enlarged at base, sometimes nearly absent . . . . (5) *R. hispidus*

    6′ Leaflets thin, dull green on upper surface; prickles stout, curved, enlarged at base. . . . . (6) *R. flagellaris*

  5′ Stems erect or arching, or occasionally trailing (blackberries):

    7 Prickles few and small or lacking; leaves glabrous or slightly hairy on veins beneath . . . . (7) *R. canadensis*

    7′ Prickles abundant, stout and broad based; leaves hairy beneath:

      8 Young growth and especially flower stalks and fruit stalks covered with many gland-tipped hairs . . . .(8) *R. allegheniensis*

      8′ Young growth and flower stalks and fruit stalks without glands or nearly so. . . . . (9) *R. argutus*

FIGURE 74 | **1. *Rubus odoratus*** flowering raspberry

An erect shrub *without prickles.* Older bark *peeling in strips.*

Leaves *simple,* 10–30 cm across, palmately veined and lobed, *maplelike,* the lobes broad based and pointed at tips, irregularly toothed, heart shaped at base; petioles and principal veins on lower side covered with purplish glandular hairs.

Stems soft-woody, light tan or straw colored, *densely bristly-hairy* with dark red or purplish gland-tipped hairs.

Buds ovate, same color as stems, with several hairy scales.

Flowers large, deep pink to purplish; June–August.

Fruits flattened, red, separating from the broad receptacle or core; edible but seedy and rather dry.

Edges of woods, moist roadsides, streambanks; occasional, throughout.

FIGURE 74 | **2. *Rubus occidentalis*** black raspberry

A shrub with *strongly arching stems.*

Leaflets three (rarely five), ovate to lanceolate, 5–8 cm long, or terminal leaflet to 10 cm long, double toothed, often with small lobes, green and glabrous above, *densely whitened with woolly hairs beneath.*

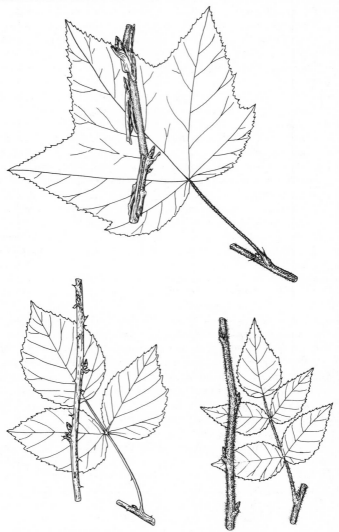

FIGURE 74   *Rubus.* Above: flowering raspberry, *R. odoratus.* Leaf, section of stem with bud and stipules, section of older stem showing buds, persistent base of leaf petioles, and shreddy bark. Below left: black raspberry, *R. occidentalis.* Leaf and stem. Below right: red raspberry, *R. idaeus* var. *canadensis.* Leaf and section of bristly stem showing lateral buds and persistent base of leaf petioles.

Young stems circular, light purple or reddish, glabrous, *covered with whitish bloom,* easily rubbed off; often rooting at tip; armed with short, broad-based prickles, these sometimes sparse.

Fruits thimble shaped, black, juicy, sweet and delicious; at maturity separating easily from the receptacle or core.

Woods, fields, thickets, fencerows; occasional, throughout.

FIGURE 74	3. *Rubus idaeus* var. *canadensis* red raspberry

Stems erect or rarely arching.

Leaflets three to seven, lanceolate to ovate, the terminal leaflet 4–10 cm long, the laterals smaller, single or double toothed, green and glabrous above, *white-woolly beneath.*

Young stems *circular* in cross section, brown to reddish, densely *glandular-bristly and covered with ashy gray wool;* prickles few or none.

Flower stalks and fruit stalks glandular and bristly.

Fruits thimble shaped, somewhat hairy, *red,* juicy, sweet and edible, readily separating from the dry receptacle or core.

Forests and clearings; throughout, common above 5,000 feet.

FIGURE 75	4. *Rubus phoenicolasius* wineberry, wine raspberry

A shrub with stems long and arching, commonly rooting at tips.

Leaflets three, the terminal 4–10 cm long, as wide as or wider than long, much larger than lateral leaflets, somewhat *heart shaped,* coarsely and irregularly toothed, often lobed; green and sparingly hairy above, *coated with whitish wool beneath;* bristly and prickly along petioles and often prickly on principal veins on lower side.

Stems rounded, reddish purple, *densely shaggy* with long, reddish brown glandular hairs or bristles and with scattered weak prickles.

Fruits nearly globose, red, sour or rather tasteless, readily separating from the dry receptacle or core.

Native of eastern Asia, well established as an escape from cultivation.

	5. *Rubus hispidus* swamp dewberry

A shrub with *trailing* or low-arching stems, usually rooting at tips and sending up short flowering stems.

Leaflets usually three, obovate, 2–7 cm long, somewhat leathery, commonly persisting through winter, single or double toothed, blunt to abruptly short pointed

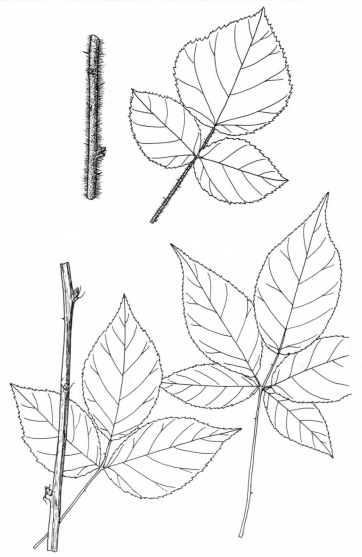

FIGURE 75   *Rubus.* Above: wineberry, *R. phoenicolasius.* Stem covered with glandular hairs and prickles, leaf with three leaflets. Below: smooth blackberry, *R. canadensis.* Stem and two forms of compound leaves.

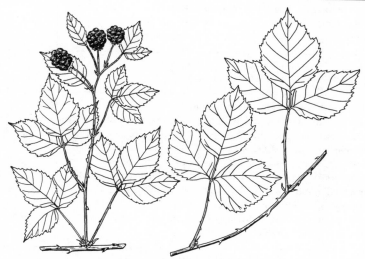

FIGURE 76   Dewberry, *Rubus flagellaris*. Left: erect stem with leaves and fruits. Right: vegetative stem with leaves.

at apex, wedge shaped to rounded at base, *lustrous, dark green above,* paler green beneath.

Stems slender, circular or angled in cross section, glandular-bristly or glabrous; prickles many to almost none, very slender, scarcely enlarged at base, straight or slightly curved.

Flower stalks and fruit stalks finely hairy, with or without stiff bristles.

Fruits small, globose, turning black slowly as they ripen, sour when mature.

Low woods, ditches, swales, wet meadows; common, throughout.

FIGURE 76 | **6. *Rubus flagellaris***
           | dewberry

A *trailing,* vinelike shrub.

Leaflets three to five, thinner than those of *R. hispidus,* ovate to lanceolate-elliptical, 2–9 cm long, coarsely single or double toothed, short pointed or long pointed at apex, wedge shaped to rounded at base, *dull green* and glabrous or finely hairy above, hairy or glabrous beneath; petioles often with a few prickles.

Stems mostly creeping over the ground and rooting at tips, but flowering and fruiting stems erect. Young stems *circular* or only slightly angled in cross section, green to reddish, glabrous or nearly so, *armed with strong prickles* curving backward (away from stem tip).

Flower stalks and fruit stalks hairy and often with glandular hairs.
Fruits large, globose to oblong, black, juicy and sweet.
Dry sandy woods, clearings, roadsides; common, NC, SC, GSMNP.

| | 7. *Rubus canadensis* |
| FIGURE 75 | smooth blackberry |

A shrub with erect or high-arching stems.

Leaves palmately compound; leaflets three to five, ovate to broadly lanceolate, 5–15 cm long, sharply toothed, long pointed at apex, rounded or heart shaped at base, *bright green and glabrous* or nearly so on both sides.

Stems *ridged or angled,* reddish purple, or reddish brown on underside, glabrous, not rooting at tips; *unarmed or with only a few weak prickles.*

Flower stalks and fruit stalks without glandular hairs.

Fruits oblong to globose, black, juicy.

Clearings and balds at higher elevations, mostly above 5,000 feet; NC; Greenville Co., SC; GA; GSMNP.

| | 8. *Rubus allegheniensis* |
| FIGURE 77 | Allegany blackberry |

A shrub with erect or high-arching stems.

Leaflets three to five, broadly ovate, 3–15 cm long, double toothed, *green on both sides,* sparingly hairy above, softly hairy beneath, velvety to touch; petioles with glandular hairs and hooked prickles.

Young stems angled or ridged, reddish to purplish, *covered with gland-tipped hairs,* at least near tip; *prickles strong,* much flattened at base, straight or curved.

Flower stalks and fruit stalks densely *glandular-hairy* and often prickly.

Fruits globose to cylindrical, black, dryish or juicy, sweet, delicious.

Open woods and thickets at high elevations; fairly common, NC, GSMNP.

| | 9. *Rubus argutus* |
| | tall blackberry, highbush blackberry |

An erect or arching shrub.

Leaflets three to five, all about alike in shape, elliptical to elliptical lanceolate, 3–12 cm long, single or double toothed, short pointed to long pointed at apex, wedge shaped or rounded at base, *green on both surfaces,* finely hairy to glabrous above, *softly hairy beneath;* prickly on petioles and principal veins.

Stems more or less hairy, at least when young, *the hairs not gland tipped;* armed with hooked, broad-based prickles. Young stems *strongly angled and grooved.*

Flower stalks and fruit stalks hairy, but the hairs usually not glandular.

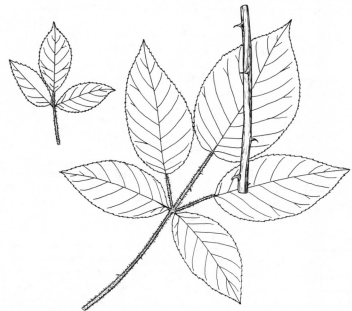

FIGURE 77   Allegany blackberry, *Rubus allegheniensis*. Two forms of compound leaves; section of stem bearing recurved prickles.

Fruit oblong, black, juicy and sweet.

Upland forests, grass balds, old fields at high elevations; common, throughout.

Rosa (rose)

Shrubs with upright, arching, or trailing stems, mostly *red* or *green* and usually *armed with prickles.*

Leaves alternate, deciduous, *pinnately compound,* the leaflets toothed.

Stipules narrow but readily seen, *attached for part of their length along base of petiole.*

Stems slender to moderate, circular in cross section; prickles commonly in pairs just below nodes.

Buds small, often set a little above leaf scar, with three or four visible scales. Terminal bud present.

Leaf scars *low, very narrow, linelike,* extending about halfway around stem. Bundle scars three. Stipule scars none.

Fruit a reddish, fleshy, berrylike aggregate called a hip, usually persistent.

1 Stipules deeply cut into narrow segments, fringelike; stems arching or
   ascending . . . . (1) R. *multiflora*
1' Stipules not deeply cut into segments:
   2 Leaflets mostly three; stems usually climbing, leaning, or trailing
     . . . . (2) R. *setigera*
   2' Leaflets five to nine; stems ascending:
     3 Lower surface of leaflets densely set with gland-tipped hairs
       . . . . (3) R. *micrantha*
     3' Lower surface of leaflets without gland-tipped hairs:
       4 Prickles slender, straight or nearly so; leaflets coarsely toothed (five
         to fifteen teeth above the middle on each side) . . . .(5) R. *carolina*
       4' Prickles stout, broad based, some or all curved or hooked down-
         ward; leaflets finely toothed (twelve to twenty-five teeth above the
         middle on each side) . . . . (4) R. *palustris*

FIGURE 78 | 1. *Rosa multiflora*
          | multiflora rose

A vigorous shrub, the *stems arching or sprawling,* commonly forming thickets.

Leaflets usually seven, obovate to elliptical, 2–4 cm long, sharply toothed, the teeth
   with red tips; short-pointed at apex, wedge shaped to rounded at base, glabrous
   above, hairy or rarely glabrous beneath; *stipules fringed or comblike;* petiole
   often prickly.

Twigs long and vigorous, flexible, red to green, glabrous, *armed with stout recurved
   prickles.*

Buds 2–3 mm long, pointed, red, glabrous, often projecting at right angles to stem;
   bud scales four or five visible.

Leaf scars long and narrow, linelike, extending about halfway around stem.

Flowers white, fragrant, numerous, in large clusters; May–June.

Fruits small, 7–9 mm across, ellipsoid to obovoid, red, glossy, smooth, *in drooping
   or erect clusters.*

Native of eastern Asia, well established as an escape from cultivation into clearings,
   roadsides, edges of woods.

---

2. *Rosa setigera*
climbing rose, prairie rose

A shrub with *long, arching, climbing, or sprawling stems.*

Leaflets mostly *three,* occasionally five, usually ovate, 3–7 cm long, sharply
   toothed, long pointed at apex, rounded at base, dark green above, paler and
   glabrous to woolly beneath; stipules with glandular hairs along margin.

Stems slender, flexible, green to reddish green, glabrous, with scattered, curved
   prickles, some of these in pairs at nodes.

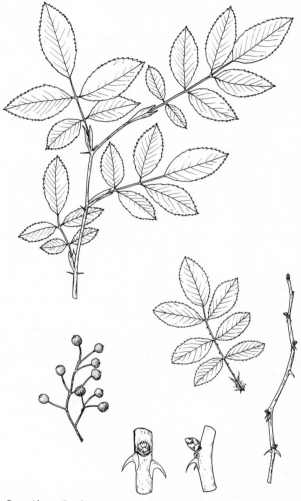

FIGURE 78   *Rosa*. Above: Carolina rose, *R. carolina*. Stem with several leaves attached, paired stipules at base of leaf petioles, and straight prickles at nodes. Below: multiflora rose, *R. multiflora*. Fruit cluster, pinnately compound leaf with stipules attached, enlarged views of lateral buds, leaf scars, and recurved prickles, and section of stem.

Buds small, ovoid, pointed, red, glabrous.
Fruits red, 8–10 mm long, in loose clusters.
Open woods, thickets, clearings; Buncombe Co., NC; GSMNP.

3. *Rosa micrantha*
sweetbrier

A shrub with many arching branches.
Leaflets five to seven, broadly ovate, 15–30 mm long, *double toothed with glandular teeth,* short pointed or long pointed at apex, wedge shaped at base, glabrous or hairy above, with *stalked glands beneath.*
Sten.s green and glabrous; armed with curved, flattened, broad-based prickles as much as 1 cm long.
Buds small, ovoid, pointed, with three or four scales visible.
Fruits 10–15 mm across, scarlet or orange.
Native of Europe, established here and there as an escape from cultivation.

4. *Rosa palustris*
swamp rose

A large shrub.
Leaflets usually seven (five to nine), narrowly elliptical to oblanceolate, 2–6 cm long, *very finely toothed,* short pointed to blunt at apex, wedge shaped or rounded at base, dull green or slightly lustrous above, paler and glabrous or finely hairy on veins beneath; *stipules usually rolled inward.*
Stems slender, green to deep red, glabrous, generally with a *pair of hooked, broad-based prickles at each node.*
Buds 2–3 mm long, red, smooth or sparingly glandular-bristly.
Streambanks and low, wet woods; occasional, throughout.

5. *Rosa carolina*
FIGURE 78 | Carolina rose, wild rose

A low shrub with erect stems.
Leaflets five to nine, narrowly elliptic to ovate, 2–6 cm long, coarsely toothed, short pointed or blunt at apex, rounded or wedge shaped at base, green and dull or slightly lustrous above, paler and glabrous beneath or hairy only on lower veins; *stipules* narrow, *flat, not fringed,* attached to petiole for nearly their entire length.
Twigs slender, green to reddish, glabrous; *prickles at the nodes slender, nearly straight, needlelike,* set at about right angles to stem; prickles or bristles often present between the nodes.
Buds 2–3 mm long, red, essentially glabrous.

Fruits about 8 mm across, nearly globose, bright red, set with glandular bristles or their scars.

Upland woods; occasional, throughout.

Prunus (cherry, plum, peach)

Trees or shrubs, many with edible fruits, some with sharp-pointed spur twigs. Older bark usually *banded crosswise with elongated lenticels.*

Leaves alternate, deciduous, simple, mostly toothed, often with raised glands at base of blade or on petiole.

Twigs slender to stout; broken twigs, especially those of cherries and peach, with characteristic *bitter almond odor and bitter taste.*

Terminal bud present in cherries and peach, absent in plums. Buds enclosed in several overlapping scales.

Leaf scars small, raised. Bundle scars three. Stipule scars present, indistinct.

Fruit a one-seeded drupe.

1 Buds woolly-hairy; twigs brightly colored, green to red, often green beneath and red on side facing sun . . . . (1) *P. persica*
1′ Buds not woolly-hairy; twigs not bright green to red:
  2 Buds clustered at tips of long, slender twigs and short spurs
  . . . . (6) *P. pensylvanica*
  2′ Buds not clustered at tips of long slender twigs; clustered buds, if present, only on spur twigs:
    3 Terminal bud absent, twig tip dying back to topmost lateral bud; twigs often thorny:
      4 Leaves mostly less than 7 cm long, folded lengthwise and appearing trough shaped; lower half of lateral buds concealed in axis between leaf petiole and twig (in the absence of leaves, lower half of buds concealed by raised leaf scar) . . . . (2) *P. angustifolia*
      4′ Leaves mostly more than 7 cm long, not folded as above; lower half of lateral buds not concealed . . . . (3) *P. americana*
    3′ Terminal bud present; twigs never thorny:
      5 Short fruiting spurs usually common on older growth, each bearing clustered buds:
        6 Leaves large, as much as 15 cm long, hairy on veins beneath; petioles variable in length but usually 2–5 cm long; fruits borne mostly on spurs without leaves . . . . (4) *P. avium*
        6′ Leaves smaller, not more than 10 cm long, glabrous beneath; petioles usually 1–2 cm long; fruits borne mostly on leafy spurs
        . . . . (5) *P. cerasus*
    5′ Short fruiting spurs absent, buds not clustered:
      7 Leaves with marginal teeth curved inward, or mostly so (magnify-

ing lens needed); lower surface of leaves often with white to rusty wool along midrib; buds reddish brown to greenish, the scales usually smooth and glossy .... (7) *P. serotina*

7' Leaves with marginal teeth pointing outward or forward, not curved inward; lower surface of leaves glabrous or with tufts of hairs in vein axils; buds dull brown or the scale edges paler brown or grayish; bud scales slightly roughened (under magnifying lens) .... (8) *P. virginiana*

1. *Prunus persica*
peach

A small tree.

Leaves lanceolate to oblong lanceolate, 7–20 cm long, finely toothed with reddish incurved teeth, *tapered to long point* at apex, wedge shaped to rounded at base, medium green above, slightly paler beneath, glabrous on both sides; petioles usually with glands at base of blade.

Twigs moderately stout, glabrous, *green to bright red, often green beneath and red on side toward light;* speckled with numerous minute lenticels; bark with odor and taste of bitter almonds.

Terminal bud present, mostly 3–4 mm long, ovoid, blunt, brown or reddish, but the scales more or less obscured by *tan or grayish wool;* buds sometimes appear clustered toward twig tip. Flower and leaf buds separate, collateral flower buds present at some nodes, at the sides of leaf buds.

Fruit a nearly globose, velvety-hairy, yellowish to reddish drupe, 3–7 cm across, with a rough stone; usually small and hard on wild plants; ripe in June or July.

Native of China, occasionally escaped from cultivation.

2. *Prunus angustifolia*
Chickasaw plum

A bushy shrub or small tree, *usually forming thickets.*

Leaves lanceolate to narrowly oblong, 2–7 cm long, *partially folded lengthwise and appearing troughlike,* evenly and very finely toothed, medium green and glossy above, paler and glabrous beneath; petioles *red,* with one or two glands near blade.

Twigs slender, zigzag, dark red to dark reddish brown, glabrous, lustrous, usually with a few thorn-tipped spur twigs.

Terminal bud absent. Lateral buds 1–2 mm long, about as thick as long, reddish brown, lustrous; *lower half of buds concealed in the axil between petiole base and twig.*

Leaf scars *much raised,* half elliptical to triangular, with a fringe of hairs along upper edge.

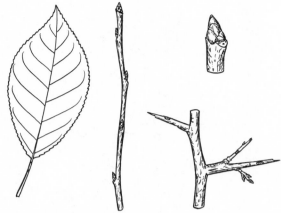

FIGURE 79   American plum, *Prunus americana*. Leaf, twig, enlarged end of twig with
              topmost lateral bud, twig scar (left) and leaf scar; section of older twig
              showing thorny lateral twigs.

Fruits 1–2 cm across, red or yellow, glaucous, ripe in late spring or early summer.
Edges of woods, roadsides, fencerows; rare in the mountains.

	3. *Prunus americana*
FIGURE 79	American plum

A slender, *thicket-forming* shrub or small tree.

Leaves obovate to oblong-ovate, 5–12 cm long, sharply single or double toothed,
the teeth spreading or forward pointing; long pointed at apex, wedge shaped to
rounded at base, dark green and glabrous above, with *sunken veins on upper
side;* glabrous beneath; petioles usually without glands, but occasionally with a
few obscure glands near base of blade.

Twigs moderately slender, reddish brown or orange brown, glabrous, entirely or
partly covered with grayish skin; broken twigs with little or no detectable odor,
but bark mildly bitter; *short growths abundant, often sharp pointed or thorny.*

Terminal bud absent. Lateral buds ovoid or broadly conical, mostly less than 4 mm
long, sharply pointed, reddish brown, glabrous; sometimes two or three buds at
node (collateral).

Leaf scars broadly crescent shaped, with *a fringe of hairs along upper edge.*

Fruits 2–3 cm across, red or yellowish, ripe in July or August.

Moist woods, roadsides, and fencerows; infrequent, throughout.

Inch plum, var. *lanata,* has densely hairy twigs and leaves. Rare in the mountains.

FIGURE 80   Mazzard, *Prunus avium*. Vegetative twig, enlarged terminal bud, leaf, section of twig bearing fruiting spurs and flower buds.

FIGURE 80	**4. *Prunus avium*** mazzard, sweet cherry

A tree, usually with well-developed trunk. Bark *smooth, glossy, reddish brown,* commonly peeling in horizontal strips.

Leaves ovate or obovate, 5–15 cm long, with rounded, usually unequal teeth, each tooth tipped with a minute dark gland; short pointed at apex, rounded or broadly wedge shaped at base, dull green and glabrous above, pale green beneath, with *hairs along lower veins;* glands present on petiole or at base of blade.

Twigs moderately slender to *stout,* pale yellowish brown to ashy gray, glabrous, glossy, more or less covered with grayish skin, easily flaked or rubbed off; bark with strong odor if bruised and with bitter taste; *short growths common on older twigs.*

Terminal bud present, 5–7 mm long, ovoid, pointed, reddish brown, glossy. Lateral buds about same size as terminal, divergent. Buds *clustered on spur twigs.*

Fruits dark red to blackish red, 10–25 mm across, the flesh firm and *sweet.*

Native of Eurasia, rarely escaped from cultivation.

	5. *Prunus cerasus* sour cherry

A small, low-branching tree.

Leaves ovate-elliptical to obovate, mostly 5–10 cm long, often smaller than those of

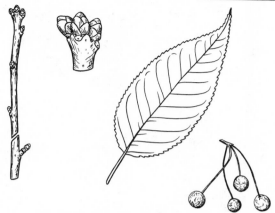

FIGURE 81   Pin cherry, *Prunus pensylvanica*. Twig; enlarged tip of twig with terminal and uppermost lateral buds and leaf scars; leaf; and fruit cluster.

P. avium, single or double toothed, the teeth blunt and gland tipped; usually short pointed at apex, rounded to broadly wedge shaped at base, dark green above, paler beneath, *glabrous* or nearly so on both sides; petioles mostly less than 2 cm long.

Twigs moderately *slender,* reddish brown, glabrous.

Buds broadly ovoid, 4–6 mm long, sharp pointed, dull brown or reddish brown, glabrous. Flower buds, all or most, *clustered* at the end of short spur twigs.

Fruits nearly globose, 15–20 mm across, bright red, juicy but *sour.*

Cultivated since ancient times, occasionally escaped.

	6. *Prunus pensylvanica*
FIGURE 81	pin cherry

A small tree with smooth, shiny, reddish brown bark.

Leaves *lance shaped,* 6–15 cm long, 2–5 cm wide, finely toothed with reddish, incurved teeth, *gradually tapering to a long, slender point* at apex, rounded at base; green, glabrous, and shiny on both sides; small glands on petiole near base of blade.

Twigs slender, *bright red* to reddish brown, glabrous, glossy, often covered with whitish skin, easily peeled or flaked off; lenticels orange colored, conspicuous on older twigs; bruised bark with strong odor and bitter taste; spur twigs common.

Buds *clustered* at tips of long twigs and short spurs. Terminal bud present, 1.5–3 mm long, ovoid to nearly globose, rounded or pointed at tip, reddish brown, glabrous, glossy; bud scales several, but difficult to see as separate scales without magnifying lens. Lateral buds in terminal cluster about same size and shape.

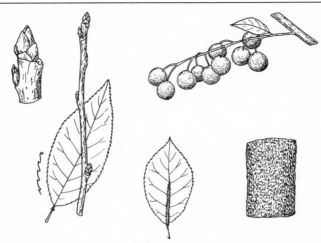

FIGURE 82   Black cherry, *Prunus serotina*. Left to right: enlarged end of twig showing
terminal bud and uppermost lateral bud; twig, upper surface of leaf, and
enlarged teeth along margin; lower surface of leaf showing rusty wool along
midrib; fruit cluster; and typical bark on young mature trunk.

Fruit a *red* cherry, 6–8 mm across, ripe in August or September.
Moist woods and cleared areas at higher elevations, mostly above 3,000 feet.

	7. *Prunus serotina*
FIGURE 82	black cherry

A medium-sized tree with narrowly spreading branches—our largest native cherry.
   Bark on young trunks and smaller branches dark red or reddish brown and
   glossy, essentially smooth but crossed by prominent, elongated lenticels. Older
   bark *blackish, scaly, the scales with upturned edges.*
Leaves lanceolate to oblong or oblanceolate, 6–12 cm long, finely toothed, *the teeth
   turning inward or pointing forward* (magnifying lens needed); short pointed or
   long pointed or rather blunt at apex, rounded to broadly wedge shaped at base;
   glossy on upper surface, pale and mostly glabrous beneath, but often with *rusty
   wool along lower midrib;* petioles generally with small glands near base of blade.
Twigs slender or moderately slender, reddish brown to greenish brown, glabrous,
   glossy, covered in places with grayish skin; yielding strong odor of bitter almonds if cut or broken; lenticels pale, dotlike, becoming elongated and prominent on older growth; *spur twigs absent.*
Terminal bud present, mostly 3–5 mm long, ovoid, bluntly pointed, uniformly reddish brown or upper scales greenish, glossy, with four to six scales visible. Lateral buds similar but slightly smaller.

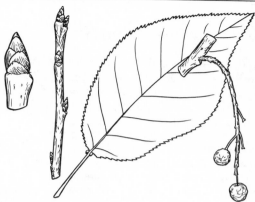

FIGURE 83   Chokecherry, *Prunus virginiana*. Enlarged terminal bud, twig, leaf, and dried remnants of fruit cluster.

Flowers small, white, numerous, in long clusters terminating short, leafy twigs; April–June.

Fruit a black cherry, 7–10 mm across, *in long clusters;* ripe in July to September.

Open woods, roadsides, fencerows, old fields; common, throughout.

FIGURE 83	8. *Prunus virginiana* chokecherry

A large, *thicket-forming* shrub or small tree. Bark *smooth* or slightly roughened, dark grayish brown.

Leaves obovate to elliptical, all or many *widest above middle,* 5–10 cm long, sharply toothed, *the teeth pointing outward or toward leaf tip, not turning inward;* short pointed or long pointed at apex, rounded to broadly wedge shaped at base; medium green and glabrous above, paler and glabrous or with tufts of hairs at juncture of veins beneath; petioles usually with two glands near blade.

Twigs moderately slender to stout, dull grayish brown to light brown, glabrous, with strong odor when freshly cut or broken, different from the cherrylike or almondlike odor of *P. serotina* and with bitter taste; *fruiting spurs lacking.*

Terminal bud present, 4–8 mm long, ovoid, sharp pointed. Bud scales six or more, keeled, dull brown with light brown or *grayish* edges and a somewhat roughened surface. Lateral buds similar.

Fruit a dark red cherry, 8–13 mm across, in long clusters; edible but more or less *puckery,* hence name chokecherry; ripe in July or August.

Rocky hillsides to wet woods; Ashe and Buncombe Cos., NC; rare in GSMNP.

Legume Family (Leguminosae)

A large family of trees, shrubs, herbs, and vines; leaves of woody members alternate, deciduous, compound except in *Cercis;* flowers regular or papilionaceous (like those in sweet pea); fruit a legume; redbud, honeylocust, black locust, and others.

FIGURE 84	*Albizia julibrissin* silktree

A low, widely spreading tree with flattened crown.

Leaves alternate, deciduous, large, *twice* pinnately compound, *fernlike or featherlike,* each leaf having six to sixteen side branchlets, each of these with eighteen to thirty pairs of *small* leaflets; leaflets lopsided, the midrib near one margin and parallel to it.

Twigs moderately slender to stout, somewhat fluted below nodes, pale green to brownish green or grayish green, glabrous; lenticels many, light colored; broken twigs with odor of green peas; pith green.

Buds superposed, the upper small but clearly visible, two- to four-scaled, glabrous, the lower buds partly or completely sunken in bark and/or concealed under leaf scar. Terminal bud absent.

Leaf scars large, raised, three-lobed, with a *smooth, shiny border.* Bundle scars three. Stipule scars small, at upper corners of leaf scar, or slender stipules often persistent.

Fruit a broad, flattened pod (legume), 10–15 cm long, containing several seeds.

Native of tropical Asia, escaped from cultivation along roadsides and edges of woods.

FIGURE 85	*Cercis canadensis* redbud

A small tree or shrub.

Leaves alternate, two-ranked, deciduous, simple (in our area the only woody legume with simple leaves), *broadly heart shaped* to nearly circular, 5–15 cm long and wide, *palmately veined,* with five to seven large veins from base, entire, dark green and glabrous above, pale and glabrous or slightly hairy beneath.

Twigs slender, zigzag, ridged downward from nodes, glabrous, dark reddish brown to purplish and dotted with pale lenticels.

Terminal bud absent. Lateral buds often superposed, reddish brown to almost black, and of two kinds (*a*) smaller leaf buds, flattened and more or less appressed, with only two scales visible, and (*b*) larger flower buds, more plump

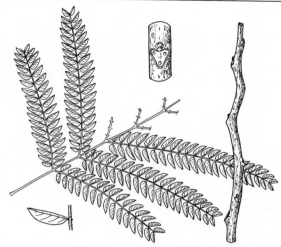

FIGURE 84   Silktree, *Albizia julibrissin*. Portion of twice-compound leaf, enlarged leaflet, leaf scar and superposed buds, twig.

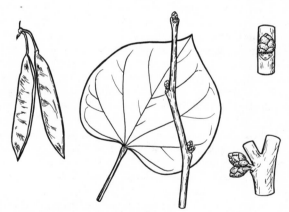

FIGURE 85   Redbud, *Cercis canadensis*. Fruits (legumes), leaf, twig, enlarged leaf bud and leaf scar, enlarged flower buds showing stalks.

and with more exposed scales, *stalked,* but the stalk concealed by bud scales, at least some of the flower buds *paired or clustered on large twigs and small branches.*

Flowers reddish or rose colored, showy, borne mostly on older stems, in spring usually before leaves.

Leaf scars alternate, two-ranked, somewhat raised, triangular in outline, with *uneven fringe along upper margin.* Bundle scars three. Stipule scars none.

Fruit a flat, thin pod (legume), 4–10 cm long, pointed at both ends, containing several hard seeds, maturing in summer or autumn and persisting well into winter.

Moist woods and ravines at lower elevations; fairly common, essentially throughout.

FIGURE 86 | *Gleditsia triacanthos*
honeylocust

A tree armed with *long, straight, usually branched thorns,* borne on trunk, branches, and twigs. Mature bark gray to almost black, relatively smooth, even on old trees, divided into long, flat plates with sharp, often curled edges.

Leaves alternate, deciduous, *once or twice pinnately compound.* Leaflets numerous, in pairs, minutely toothed or entire, 2–4 cm long on once pinnate leaves, 1–2 cm long on twice pinnate leaves.

Twigs moderately stout, conspicuously zigzag, *swollen at nodes,* greenish brown to reddish brown, smooth and polished; thorns stout, sharply pointed, three-branched (less often two-branched or single).

Terminal bud absent. Lateral buds small, several at a node (superposed), *wholly or partly concealed under leaf scar and in bark above leaf scar,* the uppermost bud commonly producing a thorn.

Leaf scars raised, unevenly shield shaped to U-shaped. Bundle scars three. Stipule scars absent.

Fruit a thin, gently curved and twisted, dark reddish brown pod (legume), 20–30 cm long, 2 cm or more across; seeds numerous, oval, somewhat flattened, surrounded by sugary pulp.

Rich, moist woods; Buncombe Co., NC; occasional in GSMNP.

FIGURE 87 | *Gymnocladus dioicus*
Kentucky coffeetree

A tree with heavy upright or spreading branches. Bark of trunk grayish brown and very rough, characterized by *thin, scalelike ridges curled up along the edges,* these hard and often sharp.

Leaves alternate, deciduous, *twice pinnately compound,* very large. Leaflets forty or more, 4–7 cm long, narrowly to broadly ovate, entire, short pointed at apex, rounded or somewhat unequal at base, dark green above, lighter green and glabrous or sparsely hairy beneath.

Twigs *very stout,* stiff and crooked, blunt at tip, brown, covered with grayish skin easily flaked off with fingernail or knife; lenticels numerous, orange colored. Pith *large, salmon pink.*

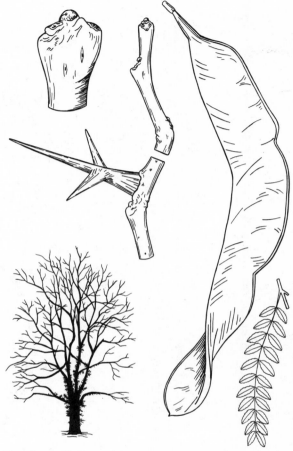

FIGURE 86   Honeylocust, *Gleditsia triacanthos*. Upper left: enlarged tip of stem showing uppermost lateral bud and stub of end of this year's growth. Center: twig and branched thorn. Right: pod (legume) and leaflets. Lower left: silhouette of mature tree.

Terminal bud absent. Lateral buds one above another (superposed), small, *silky, partly sunken in hairy craters above leaf scar.*

Leaf scars *large,* more or less heart shaped. Bundle scars three or five. Stipule scars very small or absent.

Fruit a woody pod (legume), 10–25 cm long, purplish brown, containing a few large dark brown to nearly black seeds; pods persisting on tree into winter.

Introduced in the mountains; rare.

FIGURE 87   Kentucky coffeetree, *Gymnocladus dioicus*. Detailed section of twig showing leaf scar and superposed buds, twice-compound leaf, pod (legume), and bark on mature trunk.

	Cladrastis kentukea
FIGURE 88	yellowwood

A tree with *bark smooth and gray,* similar to that of beech.

Leaves alternate, deciduous, pinnately compound. Leaflets seven to nine, large, 6–10 cm long, entire, mostly *alternate along petiole. Base of petiole hollow,* covering each bud cluster like a cone.

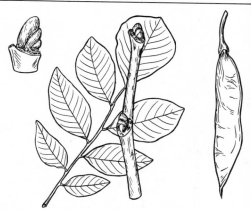

FIGURE 88 Yellowwood, *Cladrastis kentukea*. Enlarged bud cluster at end of twig and stub ending this year's growth; leaf, twig, and fruit.

Twigs moderate in size, dark brown to dark reddish brown, smooth and glossy.

Buds concealed until leaves are shed, naked, yellowish to light brown, silky-woolly, superposed (several together), crowded into a cone-shaped group and appearing as one bud. Terminal bud absent.

Leaf scars C-shaped, the open side up, *nearly surrounding the buds.* Bundle scars mostly five, often raised above surface of leaf scar. Stipules and their scars absent.

Fruit a flat, narrow pod (legume), 6–10 cm long, containing several flat seeds.

Rich woods and rocky slopes; occasional, NC, GA, GSMNP.

FIGURE 89	*Cytisus scoparius* Scotch broom

A dense shrub with upright or arching branches.

Leaves deciduous or evergreen, alternate, compound. Leaflets *three,* or upper leaves often reduced to one leaflet, obovate, 5–15 mm long, entire, short pointed at apex, wedge shaped at base, dark green and glabrous.

Twigs slender, *stiff, angled and grooved, dark green,* nearly glabrous, usually dying back from tip.

Buds small, about 1 mm long, green, glabrous, with about four scales visible, the scales often indistinct.

Leaf scars small, raised, triangular or shield shaped. Bundle scar one. Stipule scars absent or very small.

Fruit a *flat, oblong or linear pod* (legume), 3–5 cm long, hairy on edges, glabrous on both sides, containing several seeds.

Native of Europe, established in a few places as an escape from cultivation.

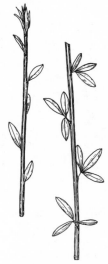

FIGURE 89   Scotch broom, *Cytisus scoparius*. Stems showing one-leaflet leaves (left) and three-leaflet leaves.

Amorpha (indigobush)

Erect shrubs; *foliage fragrant when crushed.*

Leaves alternate, deciduous, compound, with odd number of leaflets, resembling leaves of black locust (*Robinia pseudoacacia*); leaflets numerous, entire, pointed, rounded or notched at apex, *marked with small blisterlike dots on lower surface.*

Twigs rather slender, three ridged downward from nodes.

Terminal bud absent. Lateral buds small, often one above another (superposed), the upper bud largest; exposed bud scales two to four.

Leaf scars rounded-triangular to crescent shaped. Bundle scars three. Stipules slender, pointed, 2–3 mm long, persistent or leaving small scars.

Fruit a small, *warty dotted,* one- or two-seeded *pod* (legume), mostly less than 10 mm long, borne in erect, crowded clusters.

1 Stalk of leaflets distinctly hairy; midrib continuing beyond apex of leaflets as
   a bristlelike point . . . . (2) *A. fruticosa*
1′ Stalk of leaflets minutely hairy or glabrous; midrib protruding from apex of
   leaflets as a tiny, rounded knob (magnifying lens may be needed)
   . . . . (1) *A. glabra*

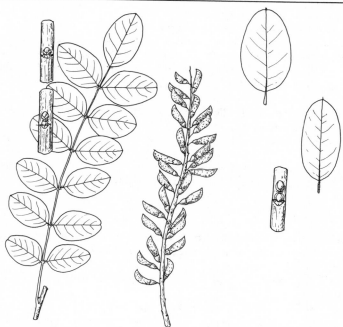

FIGURE 90   *Amorpha.* Mountain-indigo, *A. glabra.* Left, center, and upper right:
compound leaf, sections of twig showing a solitary lateral bud, superposed
buds, stipule scars, and leaf scars; an erect cluster of warty-dotted pods
(legumes), and leaflet. Lower right: false-indigo, *A. fruticosa.* Leaflet, section
of twig showing leaf scar, stipule scars, and superposed buds.

	1. *Amorpha glabra*
FIGURE 90	mountain-indigo

Leaves locustlike; leaflets nine to nineteen, mostly oblong to elliptical, 1.5–5 cm
long, entire, rounded or slightly notched at apex, *the midrib protruding as a very
small, rounded knob,* rounded or slightly heart shaped at base, *glabrous on both
surfaces,* glandular dotted beneath.

Twigs three-ridged below nodes, purplish brown to reddish or brownish purple,
glabrous or nearly so.

Buds purple to reddish purple, glabrous, two- to four-scaled, often superposed, with
one or two small buds below main one.

Stipules slender, stiff, pointed, commonly persistent; stipule scars small.

Pods mostly 7–10 mm long, warty dotted, glabrous.

Moist or dry woods, roadsides; common, essentially throughout.

FIGURE 90	2. *Amorpha fruticosa* false-indigo, indigobush

Leaves with eleven to thirty-five leaflets, oblong to elliptical, 1–5 cm long, entire, dull green and minutely hairy above, paler and usually short-hairy beneath; *midrib* (under magnifying lens) *continuing beyond apex of leaflets as a bristlelike point.*

Twigs greenish or purplish to light brown, sparingly to densely *hairy* when young, aging to nearly glabrous.

Buds flattened, appressed, often superposed (one or two smaller buds below a larger one).

Pods 6–9 mm long, distinctly warty, glabrous.

Moist woods and streambanks; occasional, throughout.

Wisteria (wisteria)

High climbing or scrambling, woody vines without tendrils or aerial rootlets.

Leaves alternate, deciduous, pinnately compound; leaflets entire.

Stems *twining,* unarmed, vigorous.

Buds solitary, appressed, with about two scales visible, the first scale nearly covering the bud.

Leaf scars much raised, often with *two small knobs or horns projecting downward* from lower side. Bundle scar *one,* in center of leaf scar. Stipules and stipule scars absent.

Fruit a flattened, beanlike pod (legume) containing several seeds.

Occasionally persisting around old homesites or escaping from cultivation.

FIGURE 91	*Wisteria sinensis* (Chinese wisteria)

This species is representative of the genus in our area. Leaflets mostly nine to eleven. Twigs, buds, and leaf petioles densely short-hairy. Native of China, occasionally escaped into roadsides and open woods.

Robinia (locust)

Rapidly growing trees and shrubs, often forming colonies by means of root sprouts.

Leaves alternate, deciduous, *pinnately compound,* with many leaflets, all but the end leaflet in opposite pairs; leaflets entire.

Twigs slender to moderately stout, zigzag, slightly to strongly angled.

Terminal bud lacking. Lateral buds small, hairy, superposed, *partly or entirely*

FIGURE 91   Chinese wisteria, *Wisteria sinensis.* Pinnately compound leaf, sections of twig showing buds and leaf scars, the lower leaf scar with woody projections pointing downward.

concealed under the leaf scars or leaf petioles; on rapidly grown twigs, an additional bud above leaf scar may develop into a rudimentary stem, soon deciduous, leaving a small branch scar.

Stipules present, *modified into spines,* often weak and pricklelike, borne *in pairs at nodes,* or sometimes lacking.

Leaf scars triangular, three-lobed, or nearly circular, ruptured as underlying buds enlarge. Bundle scars three.

Fruit a flat, many-seeded pod (legume), persisting until early winter or later.

1 Twigs and petioles densely set with stiff, bristly hairs........... (1) *R. hispida*
1′ Twigs and petioles without bristly hairs:
  2 Young twigs glandular-sticky or clammy .... (2) *R. viscosa*
  2′ Young twigs not glandular-sticky:
    3 Trees, typically fast growing, usually with strong stipular spines at the
      nodes; leaflets rounded or slightly notched at apex; spines broad based,
      somewhat triangular, very sharp, extending at wide angles on either
      side of node (where leaf is attached) ....(3) *R. pseudoacacia*

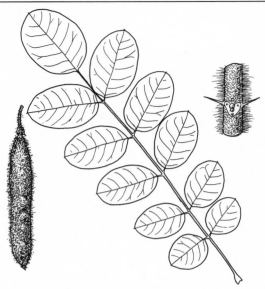

FIGURE 92	Bristly locust, *Robinia hispida*. Fruit covered with bristly hairs; pinnately compound leaf; detailed section of twig showing leaf scar, partly sunken buds, and a pair of weak stipular spines.

3' Slender shrubs 1–3 m tall; stipules slender, pricklelike or bristlelike:
    4 Leaflets oblong-lanceolate, pointed at apex .... (4) *R. kelseyi*
    4' Leaflets broader than oblong-lanceolate, blunt or rounded at apex
    .... (5) *R. boyntonii*

FIGURE 92 | **1. *Robinia hispida***
            | bristly locust

A low shrub.

Leaflets seven to nineteen, ovate-oblong to nearly circular, mostly 3–6 mm long, entire, rounded or blunt at apex and with *a short, slender, bristlelike tip* (mucro), rounded at base, bluish green and glabrous above, much paler and glabrous beneath or sparsely hairy on lower veins.

Twigs zigzag, brown, *covered with stiff bristles* and usually with a pair of weak spines at each node.

Buds superposed, partly sunken or concealed under leaf scar, but with *silky tips visible*.

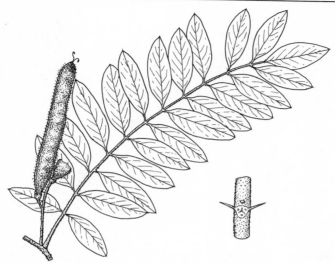

FIGURE 93   Clammy locust, *Robinia viscosa*. Fruit covered with glandular hairs, pinnately compound leaf, and section of twig showing leaf scar, superposed buds concealed under cracks in bark of leaf scar, and pricklelike stipular spines.

Leaf scars broadly triangular to three-lobed or, after being split from beneath by developing buds, deeply V-shaped or horseshoe shaped.
Pods 5–8 cm long, flat, bristly.
Open woods and roadsides; infrequent, throughout.

	2. *Robinia viscosa*
FIGURE 93	clammy locust

A shrub or small tree.
Leaflets eleven to twenty-five, ovate or ovate-oblong to elliptical, 2–5 cm long, entire, pointed to rounded at apex, the midrib continuing beyond apex as a short bristle, wedge shaped at base, dull green above, paler beneath, short-hairy or glabrous on both surfaces; *petioles with clammy glandular hairs.*
Twigs dark reddish brown, roughish warty and *covered with sticky glands;* stipules present as *weak spines or prickles.*
Buds usually concealed beneath silky surface of leaf scars.
Pods 5–10 cm long, narrowly oblong, flat, sticky with glandular hairs.
Open woods; occasional, NC.

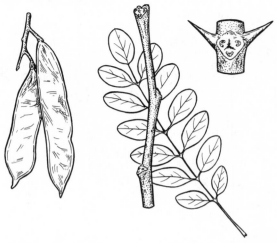

FIGURE 94   Black locust, *Robinia pseudoacacia*. Fruits, twig, pinnately compound leaf, and enlarged section of twig showing leaf scar, bundle scars, and stout stipular spines.

FIGURE 94	3. *Robinia pseudoacacia* black locust

A tall, rather narrow tree. Mature bark thick, reddish brown to nearly black, broadly ridged and *deeply furrowed*.

Leaflets seven to nineteen, elliptical to oblong-ovate, 2–6 cm long, entire, rounded to straight across or slightly notched at apex, rounded at base, bluish green and glabrous above, much paler and glabrous beneath, except for lower midrib, which may be more or less hairy.

Twigs zigzag, brittle, pale green to reddish brown, glabrous or nearly so, prominently ridged downward from sides and center of leaf scar; usually with *stout spines in pairs at nodes*.

Buds superposed, three or four together, *sunken under leaf scar,* apparently without scales, hairy.

Leaf scars irregularly shaped, often triangular or three-lobed, usually split between three bundle scars by underlying buds.

Flowers medium in size, pea shaped, white, fragrant, in drooping clusters; April–June.

Pods 5–10 cm long, flat and slightly curved, dark brown, glabrous, often persisting through winter.

Woods, thickets, old fields, roadsides; common, throughout.

4. *Robinia kelseyi*
Kelsey locust

A shrub or (rarely) a small tree.

Leaflets nine to eleven, oblong-lanceolate, 2–4 cm long, short pointed at apex, rounded at base, slightly hairy at first, soon becoming glabrous.

Twigs glabrous, with two pricklelike spines at each node.

Pods oblong, mostly 4–5 cm long, densely set with *purple glandular hairs.*

Moist or dry woods; occasional, NC, GSMNP.

5. *Robinia boyntonii*
Boynton locust

An unarmed shrub.

Leaflets seven to thirteen, elliptical to oblong, 2–4 cm long, blunt or rounded at apex, rounded at base, minutely hairy at first, soon becoming glabrous.

Twigs short-hairy or glabrous; stipular spines usually lacking, but stipules present, awl shaped, weak.

Pods rarely formed.

Moist or dry woods; occasional, NC, SC, GSMNP.

Rue Family (Rutaceae)

Shrubs or small trees with heavy-scented foliage dotted with trans-lucent glands; leaves alternate, compound, in our species decidu-ous; flowers small, in clusters; fruit a samara or capsule; prickly-ash, hoptree.

| | *Zanthoxylum americanum*
prickly-ash |

A low-branching, *aromatic* shrub *armed with stipular spines.*

Leaves alternate, deciduous, *compound.* Leaflets five to eleven, ovate, 3–6 cm long, entire or minutely toothed, short pointed at apex, rounded or broadly wedge shaped at base; upper surface dark green, *dotted with translucent glands,* visible when leaflet is held to light; lower surface paler, hairy along veins; petioles often prickly.

Twigs moderately slender, rigid, dark brown, glabrous, with *strong odor if bruised or broken,* suggesting crushed lemon peel. Spines in pairs at the nodes, resem-bling those of black locust, less than 10 mm long, stout, broad based (spines readily detachable and, botanically, may be prickles).

Buds clearly evident, globose, about 1 mm long, the scales indistinct, *red-woolly.* Terminal bud present.

Leaf scars semicircular or broadly triangular in outline. Bundle scars three. Stipule scars lacking.

Fruits small (4–5 mm long), reddish brown capsules containing one or (rarely) two seeds.

Moist woods and riverbanks; Jackson Co., NC; introduced elsewhere in the mountains.

| FIGURE 95 | *Ptelea trifoliata*
hoptree, wafer-ash |

An *aromatic* shrub or small tree.

Leaves alternate, deciduous, compound, long petioled. Leaflets *three,* 5–15 cm long, entire or finely wavy toothed, similar to leaves of poison-ivy (*Toxicoden-dron radicans*), but *terminal leaflet tapers gradually to point of attachment,* whereas terminal leaflet of poison-ivy is long stalked; dark green and lustrous on upper surface, paler and glabrous to densely hairy beneath; *covered with small, translucent dots,* visible if leaflet is held up to sunlight.

Twigs slender to moderate, brown or reddish brown, glabrous and lustrous to short-woolly, with *distinctive odor* when bruised, suggesting strong lemon peel or turpentine, and with bitter taste.

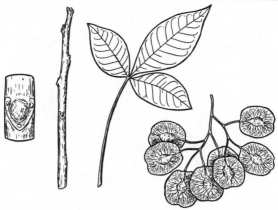

FIGURE 95   Hoptree, *Ptelea trifoliata*. Enlarged part of twig with bud and leaf scar; twig, compound leaf, and cluster of fruit.

Terminal bud absent. Lateral buds flattened, only slightly raised above surface of twig, superposed, *pale-hairy,* without apparent scales, nearly surrounded by leaf scar or commonly more or less covered by it.

Leaf scars *horseshoe shaped* or U-shaped. Bundle scars three. Stipule scars lacking.

Fruit a flat two-seeded samara with a *circular wing,* in terminal clusters, frequently persistent in winter.

Shaded streambanks and rocky slopes; scattered locations throughout.

Quassia Family (Simaroubaceae)

A family of mostly tropical and subtropical trees and shrubs; leaves alternate, pinnately compound, in our species deciduous; flowers usually in axillary clusters, staminate and pistillate generally on separate plants; fruit a samara, capsule, or drupe. Represented in our area by one species.

	Ailanthus altissima
FIGURE 96	ailanthus, tree-of-heaven

A rapidly growing tree with stout branches, often forming colonies by means of root sprouts. Bark thin, tight, medium gray to grayish brown, scored vertically with light-colored, wavy stripes. Old bark shallowly furrowed.

Leaves alternate, deciduous, pinnately compound, large. Leaflets fifteen to twenty-

FIGURE 96   Ailanthus, *Ailanthus altissima*. Detailed section of twig showing bud, leaf scar, and bundle scars; twig, compound leaf, fruit, and silhouette of mature pistillate tree.

seven, oblong to lanceolate, 7–15 cm long, the margin entire except for *a few large, gland-tipped teeth near base;* crushed leaves with strong odor, especially those from a staminate tree.

Twigs very stout and blunt, somewhat resembling those of smooth sumac (*Rhus glabra*) but *sap not milky;* yellowish brown or reddish brown to reddish green, short-velvety-hairy to nearly glabrous, with unpleasant odor when broken; pith large, brown.

Buds relatively small, solitary, appearing flattened or partly sunken in bark, with two reddish, hairy outer scales not quite meeting, an inner pair occasionally visible under outer pair. Terminal bud absent, twig tips dying back to large oval scar at side of uppermost lateral bud.

Leaf scars *large, heart shaped or shield shaped,* lighter colored than twig; margin of leaf scar raised. Bundle scars many, in a single V-shaped row near edge of leaf scar. Stipule scars absent.

Fruit a two-winged, twisted samara, 3–5 cm long, with seed cavity in center; in large clusters, produced only on plants having pistillate flowers; often persistent.

Native of eastern Asia, often escaped from cultivation into roadsides, clearings, open woods.

Mahogany Family (Meliaceae)

Shrubs and trees, mostly of tropical and warmer regions of the earth; leaves alternate, pinnately compound, in our species deciduous; flowers small, in panicles; fruit a one-seeded drupe. Represented in our area by one species.

> *Melia azedarach*
> chinaberry

A small to medium-sized tree. Bark thin and smooth, becoming furrowed with age, dark brown or reddish brown; inner bark whitish, bitter.

Leaves alternate, deciduous, *twice compound;* leaflets numerous, ovate to lanceolate-ovate, 2–8 cm long, sharply toothed, long pointed at apex, rounded at base, dark green above, paler beneath, finely hairy or glabrous on both sides; foliage with strong odor when crushed and bitter taste.

Twigs stout, circular in cross section, olive green to brownish, glabrous or nearly so; pith white.

Buds small, solitary, nearly globose, grayish brown, *appearing naked* but actually having about three indistinct scales. Terminal bud absent.

Leaf scars large, three-lobed. Bundle scars many, *in three C-shaped groups.* Stipule scars absent.

Fruits *yellow,* fleshy drupes, 10–15 mm across, in open, drooping clusters, usually persistent; flesh bitter and puckery, not edible.

Native of Asia, rarely escaped from cultivation in our area, at lower elevations.

Cashew Family (Anacardiaceae)

Deciduous trees, shrubs, or vines with milky or resinous sap; some members poisonous to touch; leaves alternate, in our species compound; flowers in conspicuous terminal or axillary clusters; fruit a small drupe; sumac, poison-ivy, poison-sumac.

Rhus (sumac)

Shrubs or small trees with *milky or resinous sap; not poisonous.*

Leaves alternate, deciduous, *pinnately compound,* made up of three to many leaflets, turning brilliant colors in autumn.

Twigs slender to stout, brittle; pith large.

Terminal bud absent. Lateral buds either *small and hairy, appearing naked, or not evident.*

Leaf scars nearly circular, broadly U-shaped or horseshoe shaped. Bundle scars several to many, scattered or in groups. Stipule scars absent.

Staminate and pistillate flowers usually on separate plants.

Fruits small, reddish, hairy drupes, borne in *terminal* clusters, often persisting through winter; with pleasant acid taste.

Summer Key

1 Leaves with three leaflets, fragrant when crushed . . . . (1) *R. aromatica*
1' Leaves with nine or more leaflets, not fragrant:
  2 Petioles winged; leaflets entire or nearly so, shiny on upper surface
  . . . . (2) *R. copallina*
  2' Petioles not winged; leaflets toothed:
    3 Twigs densely velvety-hairy. . . . . (3) *R. typhina*
    3' Twigs glabrous, usually glaucous, strongly keeled below nodes
    . . . . (4) *R. glabra*

Winter Key

1 Flower buds large, catkinlike, borne at twig tips; leaf buds concealed between twig and persistent base of leaf petiole; leaf scars circular or nearly so, at tip of petiole base . . . . (1) *R. aromatica*
1' Flower buds, leaf-buds not as above; leaf scars not circular:
  2 Twigs hairy:
    3 Hairs long, velvety, dark brown; leaf scars U-shaped or collar-shaped, almost surrounding the buds; lenticels not prominent . . . . . . (3) *R. typhina*
    3' Hairs short, fine, grayish; leaf scars about half surrounding the buds; lenticels prominent, reddish . . . . (2) *R. copallina*

FIGURE 97   Fragrant sumac, *Rhus aromatica*. Compound leaves, stem bearing catkins.

2′ Twigs glabrous, usually with whitish bloom; leaf scars almost encircling
the buds .... (4) *R. glabra*

	1. *Rhus aromatica*
FIGURE 97	fragrant sumac

A bushy shrub, usually with several stems from base, *not poisonous;* leaves and
twigs with *fragrant* odor if crushed or broken.

Leaflets three, ovate to elliptical-ovate, 2–8 cm long, rather evenly and coarsely
toothed above middle, the teeth rounded; terminal leaflet *stalkless or merely
short stalked* (or appearing short stalked because of long-tapering, wedge-
shaped base); *leaflets soft-hairy* on both surfaces, or becoming nearly glabrous.

Twigs slender, light brown to reddish brown, hairy toward tip; *aerial rootlets
absent.*

Flower buds large, noticeably scaly, catkinlike, at or near twig tips, exposed in winter. Leaf buds small, about 1 mm long, yellowish-hairy, concealed between twig
and persistent base of petiole.

Leaf scars at tip of old petiole base, nearly circular. Bundle scars five to nine.

Fruit a berrylike drupe, globose, 6–8 mm across, *bright red,* densely hairy, with a
single bony seed; borne in small, mostly terminal clusters.

Although a three-leaflet shrub, this species is readily distinguished from poison-ivy by its aromatic foliage, short-stalked terminal leaflet, hidden leaf buds, red-hairy fruits, and catkinlike flower buds.

Dry woods, rocky hillsides; Burke and Graham Cos., NC.

	2. *Rhus copallina*
FIGURE 98	shining sumac, winged sumac

A shrub or (rarely) small tree.

Leaflets nine to twenty-one (mostly nine to eleven, in pairs except the terminal leaflet), oblong to lanceolate, 4–9 cm long, *entire* or nearly so, short pointed or long pointed at apex, mostly wedge shaped at base, dark green, smooth, and shiny on upper side, pale and glabrous or hairy beneath; *petiole winged between each pair of leaflets.*

Twigs slender to moderately stout, brownish, *finely hairy;* sap watery, resinous, with odor resembling turpentine; *lenticels prominent, raised, orange red.*

Buds rounded, 1–2 mm long, brown, hairy.

Leaf scars U-shaped, about half surrounding the bud but often reaching nearly to top of bud.

Fruits 3–4 mm across, dark red, covered with short, acid-tasting red hairs, in erect terminal clusters, persistent in winter; darker red and less showy than fruits of other sumacs.

Dry woods, clearings, fencerows, thickets; common, throughout.

	3. *Rhus typhina*
FIGURE 98	staghorn sumac

A large shrub or small tree, often reaching larger size than other sumacs.

Leaflets eleven to thirty-one, lanceolate to oblong-lanceolate, 5–12 cm long, *sharply toothed,* long pointed at apex, rounded at base, green and glabrous above, paler and glabrous or somewhat hairy beneath; *petioles densely hairy, not winged.*

Twigs stout, brittle, brown to reddish, *covered with long velvety-brown hairs,* suggesting hairy antlers of young stag, the hairs concealing the lenticels; sap milky, sticky, turning black as it dries.

Buds 5–7 mm long, light brown, densely hairy, appearing naked, almost surrounded by the leaf scars.

Leaf scars large, horseshoe shaped or C-shaped, open side up, each *forming a nearly closed ring around bud.*

Fruits 3–5 mm across, bright red, covered with acid-tasting red hairs; in compact, cone-shaped clusters, persistent in winter.

Old fields, thickets, edges of woods; occasional, throughout.

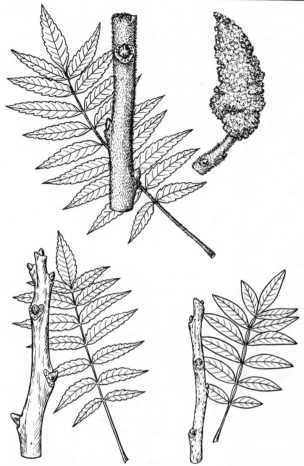

FIGURE 98   *Rhus*. Above: staghorn sumac, *R. typhina*. Leaf, section of twig, terminal fruit cluster. Lower left: smooth sumac, *R. glabra*. Leaf, section of twig. Lower right: shining sumac, *R. copallina*. Leaf, section of twig.

	4. *Rhus glabra*
FIGURE 98	smooth sumac

A large shrub or occasionally small tree.

Leaflets thirteen to thirty-one, lanceolate to narrowly oblong, 5–15 cm long, sharply toothed, long pointed at apex, rounded to wedge shaped at base, glabrous or nearly so on both sides, shiny green above, much paler beneath; petioles *glabrous,* not winged.

Twigs stout, light brown to reddish brown, *glabrous,* often with whitish bloom, *strongly keeled and three-sided below nodes;* sap milky, sticky at first but soon hardening.

Buds small, about 3 mm long, roundish, densely yellowish-hairy, apparently naked (without scales).

Leaf scars horseshoe shaped, nearly surrounding the buds.

Fruits 3–4 mm across, red, covered with short, reddish, sticky, acid-tasting hairs; crowded in terminal, pyramid-shaped clusters, persistent in winter.

Old fields, clearings, fencerows, thickets; common, throughout.

Toxicodendron (poison-ivy, poison-sumac)

Woody plants *poisonous to touch,* climbing or shrubby, or occasionally treelike.

Leaves alternate, deciduous, *compound.* Leaflets three in *T. radicans,* seven to thirteen in *T. vernix.*

Twigs slender to stout, circular in cross section, dotted with many raised lenticels; sap watery.

Terminal bud present, larger than laterals.

Leaf scars broadly crescent shaped to shield shaped, *entirely below the buds, not surrounding them,* as in most species of *Rhus.* Bundle scars numerous, in a line or scattered. Stipule scars absent.

Fruit a small *whitish* drupe, borne in loose, spreading or drooping clusters *in the axils of leaves,* persisting into winter.

1   Leaflets three, mostly ovate, coarsely toothed to entire, the terminal leaflet long stalked; stems slender; usually a climbing vine with aerial rootlets, less frequently trailing or shrubby; common species ....(1) *T. radicans*

1′ Leaflets seven to thirteen, oblong or lanceolate, entire; stems stout; aerial rootlets absent; a coarse shrub or small tree of wet places......... (2) *T. vernix*

FIGURE 99 | **1. *Toxicodendron radicans***
          | poison-ivy

A woody vine climbing by *aerial rootlets,* or sometimes a low, erect shrub; *poisonous to touch.*

Leaflets three, ovate to elliptical, mostly 4–14 cm long, toothed with a few irregular teeth or less commonly entire, long pointed or short pointed at apex, rounded at base, medium green and dull or glossy above, paler and often slightly hairy beneath; *terminal leaflet long stalked,* lateral leaflets nearly without stalk.

Twigs moderately slender, yellowish brown to greenish brown, slightly hairy or glabrous; lenticels usually conspicuous; aerial rootlets commonly lacking where plant is not climbing.

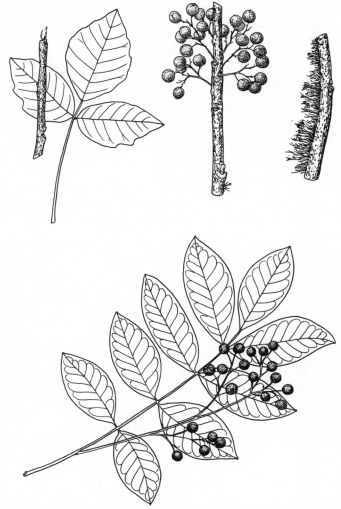

FIGURE 99   *Toxicodendron*. Above: poison-ivy, *T. radicans*. Leaf, stem, stem showing axillary fruit clusters, stem showing aerial rootlets. Below: poison-sumac, *T. vernix*. Leaf and fruit clusters.

Terminal bud 4–6 mm long, *naked,* flattened, slender, coated with tan or light reddish brown hairs. Lateral buds much smaller, often stalked.

Leaf scars large, broadly V-shaped or crescent shaped, with several bundle scars. Stipule scars none.

Fruit a small, 4–5 mm across, roundish, *white or gray,* berrylike drupe, *borne in loose clusters in the axils of leaves,* commonly lasting into winter.

Open woods, thickets, old fields, fencerows, roadsides; common, throughout.

| | 2. *Toxicodendron vernix* |
| FIGURE 99 | poison-sumac |

A loosely branched shrub or occasionally a small tree. Sap *highly poisonous to skin,* even in winter. Bark grayish and smooth.

Leaflets seven to thirteen, oblong to lanceolate, 5–13 cm long, *entire,* short pointed to long pointed at apex, wedge shaped at base, dark green and glabrous above, paler and glabrous beneath; resembling leaves of white ash (*Fraxinus americana*) but *alternate.*

Twigs moderate to stout, reddish brown or yellowish brown aging to gray, *glabrous,* glaucous; speckled abundantly with raised, dotlike lenticels; sap clear, watery, *very toxic,* turning black as it dries. Pith large, white, circular in cross section.

Terminal bud present, 3–6 mm long, occasionally longer, at least twice as large as the laterals, conical, the scales overlapping, purple, finely hairy to glabrous. Lateral buds globose.

Leaf scars relatively large, broadly crescent shaped, rounded-triangular, or shield shaped. Bundle scars many, scattered.

Fruit a small, 5–7 mm across, waxy, *white or whitish,* berrylike drupe, similar to that of *T. radicans,* borne in *open, drooping,* grapelike *clusters* in axils of leaves near end of twigs, never in dense upright clusters as in the harmless sumacs (*Rhus*); may be persistent through winter.

Wet or marshy ground; southwest mountains of NC; GA.

Holly Family (Aquifoliaceae)

Trees or shrubs with watery sap; leaves alternate, simple, evergreen or deciduous, usually toothed, the teeth sometimes spiny; flowers small (in our species), white or greenish, in axillary clusters, mostly unisexual, staminate and pistillate flowers on separate plants; fruit a small drupe. The family is represented in North America by only one genus.

Ilex (holly)

Shrubs or small trees, deciduous or evergreen.

Leaves alternate, simple, toothed.

Twigs slender to stout, circular in cross section.

Buds commonly superposed, the upper one larger, mostly with two to six scales. Terminal bud usually present.

Leaf scars half circular to crescent shaped. Bundle scar only *one. Stipules often persistent, very small, triangular, pointed, brown to black,* or their minute scars present, visible under magnifying lens.

Staminate and pistillate flowers generally on separate plants. Fruit a small *berry-like drupe,* containing four to six seedlike nutlets, borne in clusters in axils of leaves, sometimes persistent.

1 Leaves thick and leathery, persistent, usually with sharp, spiny teeth
   .... (1) *I. opaca*
1′ Leaves not thick and leathery, deciduous:
   2 Slow-growing spur twigs numerous after first year; buds pointed, the lateral buds appressed or set at narrow angle to twig; nutlets ribbed:
      3 Leaves glabrous or nearly so..........................(2) *I. montana*
      3′ Leaves soft-hairy beneath....................(2) *I. montana* var. *mollis*
   2′ Slow-growing spur twigs few or none; buds blunt, the lateral buds set at wide angle to twig; nutlets smooth........................ (3) *I. verticillata*

FIGURE 100	1. *Ilex opaca* American holly

An *evergreen* tree. Bark of trunk light gray, smooth or roughened by warty outgrowths.

Leaves *stiff and leathery,* elliptical, 5–10 cm long, *spiny tipped and usually spiny margined,* wedge shaped at base, dark green and either dull or somewhat lustrous on upper side, paler beneath.

Twigs moderate to stout, rusty-hairy at first, soon glabrous and pale brown.

FIGURE 100   *Ilex.* Above: American holly, *I. opaca.* Leaves and fruits. Lower left: mountain winterberry, *I. montana.* Leaves, twig showing lateral spur twigs on older growth, and section of twig showing leaf scar, persistent stipules, and superposed buds. Lower right: common winterberry, *I. verticillata.* Leaf, twig, and section of twig showing leaf scar and superposed buds.

Terminal bud present, 2–4 mm long, scaly, hairy.

Fruits orange to red, 7–10 mm across, borne singly or in clusters of two or three in axils of leaves or just below leaves, persistent through winter.

Moist woods at low and middle elevations; common, throughout.

| | 2. *Ilex montana* |
| FIGURE 100 | mountain winterberry |

A large shrub or small tree.

Leaves *deciduous,* alternate, often clustered at tips of short spurs, thin, oblong-lanceolate to ovate, 6–12 cm long, finely and sharply toothed, the teeth pointing forward, long pointed at apex, wedge shaped at base, dull green above, paler and glossy beneath, glabrous on both sides (petioles and undersurface of leaves softly hairy in var. *mollis*).

Twigs slender to moderate, often zigzag, purplish or greenish to reddish brown, glabrous. *Spur twigs common on older growth,* each bearing crowded leaf scars and, in season, a few leaves.

Terminal bud present, at least on some twigs. Bud scales keeled, pointed, chestnut brown to purplish brown, fringed with hairs but otherwise glabrous. Lateral buds appressed or narrowly spreading, commonly superposed, the upper bud larger, broadly ovoid, with about six scales visible, the lower bud smaller, with two scales showing.

Fruits red, about 1 cm long, borne on very short stalks, singly or in groups of two or three, often on short spurs; *nutlets grooved.*

Rich woods and mountain slopes; common, throughout.

| | 3. *Ilex verticillata* |
| FIGURE 100 | common winterberry |

A large shrub.

Leaves deciduous, ovate, obovate, or oblong-lanceolate, 4–10 cm long but mostly 6 cm long or less, sharply and finely toothed, long pointed at apex, wedge shaped at base, *dull green above,* paler and hairy beneath; veins appear sunken in upper surface.

Twigs very slender, lined downward from nodes, purplish or greenish to gray, or commonly red where exposed to sunlight, glabrous to hairy; *spur twigs generally lacking.*

Buds blunt, 1–2 mm long, *pointing outward at wide angle to stem,* brown, glabrous, often superposed, a small bud at the base of a larger one.

Fruits bright red, 5–7 mm long, on very short stalks, solitary or in crowded clusters of a few, in axils of leaves, persistent after leaves have fallen; *nutlets smooth.*

Low woods, wet thickets, streambanks; occasional, throughout.

Staff-tree Family (Celastraceae)

Shrubs, rarely treelike, or twining woody vines; leaves opposite or alternate, simple, toothed, in our species deciduous; flowers small, in clusters, bisexual or staminate and pistillate separate; fruit (in our members) a leathery capsule, the seeds usually with a brightly colored covering; euonymus, bittersweet.

Euonymus (euonymus)

Erect or trailing shrubs with *opposite* leaves and *greenish, four-sided stems.*
Leaves deciduous, simple, toothed.

Twigs with four lines or ridges extending downward from each pair of lateral buds; pith greenish.

Terminal bud present, covered by three to five pairs of overlapping scales. Lateral buds similar but smaller.

Leaf scars opposite or occasionally subopposite, only slightly raised, half circular to crescent shaped; opposing leaf scars not meeting and not joined by a transverse line. Bundle scar one, dotlike or a short curved line. Stipule scars very small, often indistinct and difficult to see without magnifying lens.

Fruit a brightly colored capsule, splitting open in autumn and disclosing *seeds with orange red covering;* persistent into winter.

1 Leaves with petioles less than 5 mm long; twigs more or less four-sided in cross section:
  2 Plant erect and self-supporting, with stiff, widely spreading branches; uppermost leaves widest at or below the middle, two to four times as long as wide, nearly without a petiole...........................(1) *E. americanus*
  2′ Plant trailing, rooting at nodes, with a few ascending branches; uppermost leaves widest at or above the middle, often less than twice as long as wide, with a distinct petiole ....(2) *E. obovatus*
1′ Leaves with petioles 10 mm long or more; twigs roundish in cross section but four-lined below nodes............................ (3) *E. atropurpureus*

FIGURE 101 | 1. *Euonymus americanus*
           | strawberry-bush

An erect or ascending shrub.
Leaves *lanceolate to narrowly ovate,* 3–9 cm long, finely toothed with incurved teeth, short pointed to long pointed at apex, wedge shaped to rounded at base; both surfaces bright green, glabrous or nearly so; *petioles 1–3 mm long.*

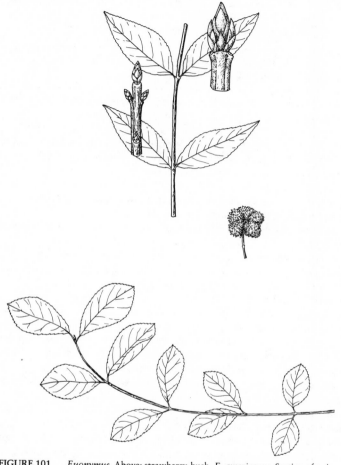

FIGURE 101   *Euonymus*. Above: strawberry-bush, *E. americanus*. Section of twig
showing opposite buds and leaf scars; twig and leaves; enlarged terminal
bud and topmost lateral buds; and fruit. Below: running strawberry-bush, *E.
obovatus*. Trailing stem and leaves.

Twigs slender to moderate, four-ridged downward from nodes, thus more or less
four-sided, glabrous, *green;* lateral twigs stiffly divergent from main stem; pith
spongy, light green.

Terminal bud narrowly oblong-ovoid, 3–8 mm long, pointed, with four or five pairs
of scales in four ranks; bud scales keeled, long pointed, green tinged with red or
brown. Lateral buds similar but smaller.

Fruit crimson when ripe, rough-warty, flattened, three- to five-lobed, containing
  scarlet seeds; persistent into winter.
Moist or dry woods; fairly common, throughout.

FIGURE 101 | 2. *Euonymus obovatus*
running strawberry-bush

A low shrub, the main stems *trailing,* often taking root and sending up a few stems
  5–30 cm high.
Terminal pair of leaves *obovate,* 3–9 cm long, finely toothed, *abruptly tapering to
  a blunt point* at apex, wedge shaped at base; lower leaves usually narrower and
  smaller; dull light green above, green beneath, glabrous on both surfaces or
  sometimes hairy on lower veins; *petioles mostly 3–4 mm long.*
Twigs slender, four-sided, green, glabrous or rarely hairy.
Terminal bud narrowly ovoid, pointed, greenish, with three or four pairs of scales
  exposed. Lateral buds small, often not evident.
Fruits rough-warty capsules, orange red, generally three-lobed, containing scarlet-
  coated seeds.
Streambanks, moist woods; occasional, throughout.

---

3. *Euonymus atropurpureus*
burningbush, wahoo

An erect shrub or rarely a small tree.
Leaves opposite, elliptical to ovate, 6–12 cm long, finely toothed, short pointed to
  long pointed at apex, broadly wedge shaped at base, dark green on both sides,
  glabrous above, hairy beneath; *petioles 10–15 mm long.*
Twigs slender, stiff, *green* to dark purplish brown, glabrous, nearly circular in cross
  section but frequently marked with *four longitudinal lines;* pith spongy.
Terminal bud conical, 3–5 mm long, purplish, glabrous, with three to five pairs on
  scales visible. Lateral buds small, greenish.
Fruit about 2 cm across, reddish or purplish, deeply four-lobed, each lobe splitting
  open and exposing scarlet covering around seed; in drooping clusters, usually
  remaining on plant until midwinter.
Streambanks and moist, rich woods; Cherokee Co., NC; rare, GSMNP.

Celastrus (bittersweet)

Vines without tendrils or, where support for climbing is lacking, sometimes low,
single-stemmed shrubs.
  Leaves alternate, deciduous, simple, thin, sharply or bluntly toothed.
  Stems *twining,* moderately slender, circular in cross section.
  Buds small, *projecting outward at nearly right angle to stem,* with six or more
scales, the outer scales long pointed. Terminal bud absent.

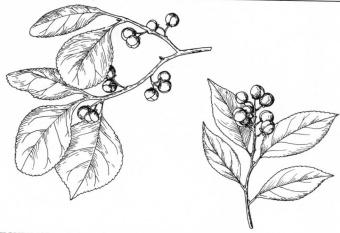

FIGURE 102   *Celastrus.* Left: Oriental bittersweet, *C. orbiculatus.* Stem, leaves, and axillary groups of fruits. Right: American bittersweet, *C. scandens.* Stem, leaves, and terminal cluster of fruit.

Leaf scars slightly raised, semicircular or half elliptical. Bundle scar one, transverse or C-shaped, open side up. Stipule scars minute and barely distinguishable, or stipules persistent, resembling tufts of hairs.

Fruit a globose, *orange or orange yellow capsule,* in terminal clusters or in small axillary groups, opening in late summer or early autumn and *exposing fleshy reddish covering around seeds;* fruits often persistent through winter.

1 Leaves of lateral stems mostly about twice as long as wide; flowers and fruits in terminal clusters, five to twenty-five in cluster . . . . (2) *C. scandens*
1' Leaves of lateral stems mostly less than twice as long as wide; flowers and fruits in axillary clusters, usually two or three in cluster . . . . . . (1) *C. orbiculatus*

FIGURE 102	**1. *Celastrus orbiculatus*** Oriental bittersweet

Leaves *nearly circular* to broadly ovate when mature, 4–7 cm long, finely and bluntly toothed, short pointed or abruptly long pointed at apex, wedge shaped at base, green and glabrous on both sides.

Stems climbing or twining, rather slender, brown to gray, glabrous.

Buds *pointing outward* at wide angle to stem, about 2 mm long, brown, glabrous.

Fruits about 1 cm across, *in small clusters in axils of leaves.*

Native of eastern Asia, established as an escape, common in a few places.

<table>
<tr><td>FIGURE 102</td><td>2. Celastrus scandens
American bittersweet</td></tr>
</table>

Leaves *ovate-oblong* when mature, 4–10 cm long, finely toothed, long pointed at apex, wedge shaped at base, medium green above, paler beneath, glabrous on both sides.

Stems brown and glabrous.

Buds small, globose, covered with several long-pointed scales.

Fruits 10–13 mm across, *in clusters terminating stems of the season.*

Thickets, woods; occasional, essentially throughout.

Bladdernut Family (Staphylaceae)

Shrubs or trees with mostly opposite, deciduous, compound leaves with three or more leaflets; flowers in terminal or axillary clusters; fruit a bladdery capsule enclosing one or more seeds. Represented in our area by one species.

| | *Staphylea trifolia*
bladdernut

A shrub or small tree. Older bark greenish or grayish *striped* with white.

Leaves opposite, deciduous, compound. Leaflets *three* (rarely five), elliptical to ovate, mostly 4–8 cm long, finely toothed, long pointed at apex, broadly wedge shaped or rounded at base, green above, paler and hairy beneath; terminal leaflet long stalked.

Twigs slender, rounded, green to brown or yellowish, glabrous, lustrous.

Buds small, 3–5 mm long, pointed, reddish brown or brown, mostly glabrous, with usually four scales visible. Terminal bud absent, the topmost lateral buds paired at twig tip.

Leaf scars opposite, small, triangular to broadly crescent shaped; opposing scars not connected by lines. Bundle scars three or more. Stipule scars *present, distinct.*

Fruit a three-celled, *bladdery, podlike capsule,* 3–5 cm long at maturity, containing a few rounded, shiny seeds, ripe in late summer.

Moist woods and shaded streambanks; Madison and Polk Cos., NC; Pickens Co., SC; rare, GSMNP.

Maple Family (Aceraceae)

Trees or shrubs with sweet, watery sap; leaves opposite, deciduous, mostly simple and palmately lobed (compound in *Acer negundo*); flowers usually greenish, small, and clustered; fruit consisting of two long-winged samaras, united at base, each one-seeded and known as a *key.*

Acer (maple)

Deciduous trees or shrubs.

Leaves *opposite,* mostly simple (pinnately compound in *A. negundo*), *palmately lobed* and veined.

Twigs moderately stout to slender, rounded, often highly colored.

Terminal bud present, with either valvate or overlapping scales. Lateral buds solitary or collateral, sometimes stalked.

Leaf scars opposite, narrow, curved, more or less U-shaped; *opposing leaf scars meeting around twig or connected by a transverse ridge.*

Bundle scars typically *three,* sometimes more. Stipule scars none.

Fruit a two-winged *double samara,* separating at maturity, each part one-seeded.

Summer Key

1 Leaves compound, consisting of three to five (rarely seven) leaflets, their margins coarsely toothed or slightly lobed...................... (1) *A. negundo*
1′ Leaves simple, lobed:
  2 Leaf lobes with a few large teeth but not consistently toothed:
    3 Leaves pale green and mostly glabrous beneath; petioles of terminal pair of leaves not greatly enlarged, the lateral buds clearly evident; stipules absent.. (2) *A. saccharum*
    3′ Leaves yellowish green and softly hairy beneath, the sides usually drooping; petioles of terminal pair of leaves greatly enlarged at base, nearly or entirely concealing lateral buds, often with leaflike stipules attached.. (3) *A. nigrum*
  2′ Leaf lobes either irregularly toothed or finely to coarsely toothed:
    4 Leaves finely toothed, with six to ten teeth per centimeter of margin; older twigs striped with white or light green........ (4) *A. pensylvanicum*
    4′ Leaves not so finely toothed, with fewer than six teeth per centimeter of margin; older twigs not striped with white or green:
      5 Leaves three- to five-lobed, the terminal lobe less than half the length of blade:
        6 Twigs pubescent; lower surface of leaves not whitened
          ....(5) *A. spicatum*
        6′ Twigs glabrous; lower surface of leaves usually whitened
          .... (6) *A. rubrum*
      5′ Leaves deeply five-lobed, the terminal lobe more than half the length of blade....................................... (7) *A. saccharinum*

Winter Key

1 Opposite leaf scars meeting around twig and prolonged upward (toward twig tip) into points; buds blue or purplish to deep red, more or less woolly
  .... (1) *A. negundo*
1′ Opposite leaf scars not meeting around twig or, if meeting, not prolonged upward into points:
  2 Buds with one pair of outer scales as long as the bud:
    3 Terminal buds 8–12 mm long, glabrous; older bark streaked with white or pale green ....(4) *A. pensylvanicum*

3' Terminal buds 5–7 mm long, short-hairy; older bark not streaked
. . . .(5) *A. spicatum*
2' Buds with several to many overlapping scales, the outer pair shorter than the bud:
  4 Terminal buds ovoid, blunt, with two to four pairs of scales visible; bud scales reddish or orange; clusters of roundish flower buds often present:
    5 Twigs with unpleasant odor if cut or broken; bark on young trees silvery, on older trunks flaking off in thin pieces . . . . .(7) *A. saccharinum*
    5' Twigs nearly odorless if cut or broken; bark on young trunks smooth and light gray, as in beech, later rough ridged but seldom flaky
. . . . (6) *A. rubrum*
  4' Terminal buds cone shaped, sharply pointed, with mostly six pairs of scales visible; bud scales brown or blackish; clustered flower buds never present:
    6 Twigs slender, reddish brown; buds glabrous or hairy only toward tip and along edge of scales; common at higher elevations
. . . . (2) *A. saccharum*
    6' Twigs moderately stout, yellowish brown to orange brown; buds hairy over all or most of their surface; species rare in mountains
. . . . (3) *A. nigrum*

FIGURE 103 | 1. *Acer negundo*
boxelder, ashleaf maple

A small to medium sized tree.

Leaves *compound;* leaflets usually three to five, occasionally seven to nine, ovate or elliptical, 7–12 cm long, irregularly and coarsely toothed or slightly lobed, long pointed at apex, rounded to wedge shaped at base, light green and glabrous above, paler and usually hairy beneath.

Twigs moderate in size, *bright green,* or purplish where exposed to sun, glabrous, usually glaucous.

Buds more or less *woolly with whitish hairs,* blue, purplish, or deep red beneath the wool; bud scales opposite in pairs, one or two pairs showing, the outer pair joined at base; terminal bud bluntly pointed or rounded, closely flanked by uppermost pair of lateral buds; lateral buds plump, stalkless or short stalked, often nearly glabrous.

Leaf scars opposite, V-shaped, each pair of scars meeting around twig and *extended upward into a sharp angle.*

Fruits in long, drooping clusters; maturing by autumn.

Low woods and streambanks; infrequent, throughout.

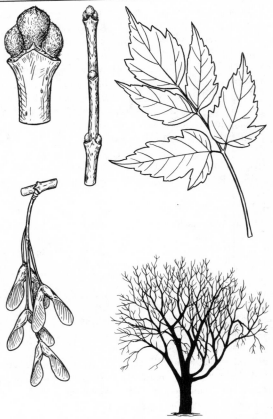

FIGURE 103   Boxelder, *Acer negundo*. Enlarged end of twig with terminal bud and uppermost pair of lateral buds; section of twig; compound leaf, fruit cluster, and silhouette of mature tree.

FIGURE 104 | 2. *Acer saccharum*
sugar maple

A large tree. Mature bark variable; grayish or brownish, rather smooth or rough, with broad, platelike ridges, the ridges sometimes loose along one edge, or with deep furrows and narrow, scaly ridges.

Leaves 7–15 cm long, five-lobed (rarely three-lobed), lobes sharply pointed and with *a few large teeth;* heart shaped at base; green and glabrous above, paler and usually glabrous beneath; petioles slender, glabrous, only slightly enlarged at base, not greatly enlarged as in *A. nigrum;* stipules absent.

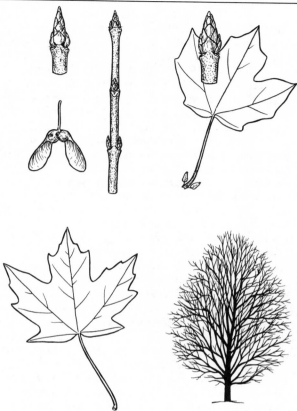

FIGURE 104   *Acer.* Upper left and below: sugar maple, *A. saccharum.* Terminal bud and
topmost lateral buds; paired fruits; section of twig showing buds and leaf
scars; typical leaf; silhouette of mature tree. Upper right: black maple, *A.
nigrum.* Buds at end of twig, leaf showing leafy appendages at base of
petiole.

Twigs slender, greenish brown or reddish brown to brown, glossy; lenticels whitish,
    conspicuous.
Terminal bud 5–7 mm long, *conical, sharply pointed,* brown, glabrous or short-
    hairy near tip, with *many scales* (four to eight pairs) *visible.* Lateral buds similar
    but smaller, rather closely appressed, with about four pairs of scales exposed.
Fruit with wings moderately spreading, glabrous; maturing in late summer or
    autumn.
Rich woods; common in NC and GSMNP, especially below 5,000 feet.

	3. *Acer nigrum*
FIGURE 104	black maple

A large tree. Bark of trunk *darker, more deeply furrowed* than that of *A. saccharum.*

Leaves 7–15 cm long, 7–16 cm wide, usually *three-lobed and drooping at sides,* the lobes entire or with a few wavy teeth; long pointed at apex, heart shaped at base, dull green above, *yellowish green* and more or less hairy beneath; petioles *stout,* thickened at base and usually enclosing lateral buds, often with two stipules or leaflike appendages attached.

Twigs moderate, brown to yellowish brown or orange brown, hairy when young, later glabrous; *lenticels prominent,* warty, light colored; clusters of bud scale scars mark beginning of each season's growth.

Terminal bud 4–7 mm long, conical, sharp pointed, dark brown to nearly black, more or less evenly covered with short hairs, with about six pairs of overlapping scales. Lateral buds similar but smaller, usually short stalked, partly or entirely concealed by clasping base of leaf petioles.

Fruit with wings divergent, somewhat U-shaped.

Moist woods; Ashe, Madison, Swain, Yancey Cos., NC.

	4. *Acer pensylvanicum*
FIGURE 105	striped maple

A large shrub or small tree. Bark of trunk and larger branches greenish *striped with white.*

Leaves larger than those of *A. spicatum,* 10–20 cm long and nearly as wide, mostly three-lobed, the middle lobe largest, the side lobes pointing forward, the margin *finely and sharply toothed;* rounded to heart shaped at base; dark green and glabrous above, slightly paler and mostly glabrous beneath.

Twigs moderate, purplish red to purplish green or reddish green, *glabrous,* covered in places with whitish bloom; each year's growth marked at base by two circular scars left by fallen bud scales.

Terminal bud present, narrowly ovoid, prominently long stalked, 8 mm or more long, including stalk, *blunt,* deep red, *glabrous,* glossy; visible bud scales two, valvate, sharply keeled, the keel extending down the stalk. Lateral buds smaller, appressed.

Fruit with widely spreading wings.

Rich, cool woods; common, throughout.

	5. *Acer spicatum*
FIGURE 105	mountain maple

A large shrub or small tree. Bark *not striped* as in *A. pensylvanicum.*

Leaves simple, 7–12 cm long and wide, usually three-lobed, occasionally five-lobed,

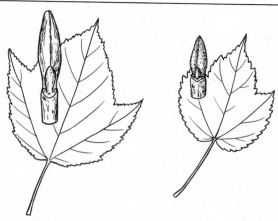

FIGURE 105   *Acer.* Left: striped maple, *A. pensylvanicum.* Leaf, enlarged terminal bud and uppermost lateral bud. Right: mountain maple, *A. spicatum.* Leaf, enlarged terminal bud and uppermost lateral bud.

the lobes long pointed, *coarsely toothed,* mostly deeply heart shaped at base, yellowish green and glabrous above, paler and pubescent beneath; long petioled.

Terminal bud narrowly ovoid, *pointed* at tip, short stalked, 5–7 mm long including stalk, red or greenish, more or less *grayish-hairy;* bud scales strongly keeled, the outer pair *valvate,* but sometimes spreading at tip to expose second pair of hairy scales. Lateral buds similar but smaller, flattened, somewhat appressed, stalked.

Fruit with wings short (10–18 mm long) and spreading at acute or nearly right angle; maturing in late summer or autumn.

Rich, cool woods; common at higher elevations, mostly above 3,000 feet; NC, GA, GSMNP.

	6. *Acer rubrum*
FIGURE 106	red maple

A medium-sized to large tree. Bark on young trunks and larger branches *smooth, gray;* becoming dark and scaly ridged on older trunks (scales do not flake off as in *A. saccharinum*).

Leaves simple, 5–12 cm long, three- to five-lobed, with *open, V-shaped, angular sinuses* between lobes, the middle lobe usually less than half the length of the blade, coarsely toothed, green and glabrous above, *whitish* or greenish and usually glabrous beneath; petioles reddish.

Twigs slender, *bright red* to dark brownish red, *glabrous,* odorless or nearly so when broken.

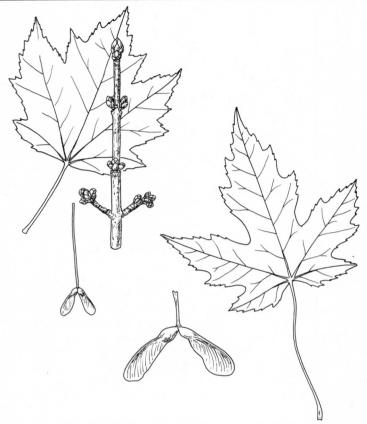

FIGURE 106	*Acer.* Left: red maple, *A. rubrum.* Leaf; twig showing terminal bud, lateral leaf buds, and clustered flower buds; fruit. Right: silver maple, *A. saccharinum.* Fruit and leaf.

Terminal bud broadly ovoid, 4–5 mm long, bluntly pointed, red, with about three (two to four) pairs of scales visible, the scales glabrous except for hairs along margin. Lateral leaf buds very short stalked, with two or three pairs of scales exposed. Flower buds usually numerous, globose, *clustered* at nodes, conspicuous.

Fruits with wings slightly spreading; maturing and dropping in spring.

Low woods and moist uplands; our most common maple, throughout.

| FIGURE 106 | 7. *Acer saccharinum*
silver maple |

A large tree. Bark at first silvery gray and smooth, eventually dark gray, rough and shaggy, often *scaling off and leaving brown spots.*

Leaves 8–15 cm long, deeply five-lobed, with narrow or subacute sinuses between lobes, *the middle lobe more than half the length of the blade;* the margin of lobes again lobed and irregularly toothed; bright green and glabrous above, *silvery* or very pale beneath.

Twigs slender, flexible, *reddish brown* to red or light brown, glabrous; *giving off unpleasant odor* when broken or bruised.

Terminal bud 3–4 mm long, broadly ovoid, bluntly pointed, with two to four pairs of scales showing, the scales reddish, slightly keeled, margined with hairs but otherwise glabrous.

Fruits with wings as much as 5 cm long and widely spreading; maturing and dropping in spring.

Bottomlands and riverbanks; occasional, throughout.

Horse-chestnut Family (Hippocastanaceae)

Trees or shrubs with a large terminal bud and opposite, deciduous, palmately compound leaves; leaflets toothed and straight veined; flowers in showy terminal clusters; fruit a rather leathery, thick-walled, mostly one-seeded capsule.

Aesculus (buckeye)

Trees or shrubs.

Leaves *opposite,* deciduous, *palmately compound,* made up of five to seven toothed leaflets, with long petiole.

Twigs stout, circular in cross section, sometimes ending in flower scar or fruit scar.

Terminal bud present; buds *large and scaly,* covered with many overlapping scales.

Leaf scars opposite, large, conspicuous, shield shaped or triangular, each pair connected by a transverse ridge. Bundle scars five or more, arranged in a V-shaped line or in three groups. Stipule scars absent.

Fruit a large, roundish *capsule* with thick, leathery husk, enclosing one to three large, tan to dark brown, nutlike seeds, each having a *light brown circular scar.*

1 Terminal buds very large, mostly 15 mm long or more; fruits 5–8 cm across, the husk 4–10 mm thick....................................(1) *A. octandra*
1' Terminal buds smaller, mostly less than 12 mm long; fruits 2–4 cm across, the husk 2–3 mm thick.....................................(2) *A. sylvatica*

| FIGURE 107 | 1. *Aesculus octandra* yellow buckeye |

A medium-sized to large tree. Mature bark thin, dark brown, roughened by shallow furrows and loose scales.

Leaflets five, rarely seven, lanceolate to obovate, 10–22 cm long, finely and somewhat irregularly toothed, long pointed at apex, wedge shaped at base, green and glabrous above, paler and hairy to nearly glabrous beneath.

Twigs light brown, glabrous; lenticels numerous; many bud-scale scars at base of each season's growth.

Terminal bud 15–20 mm long, conical, pointed, pale brown to reddish brown, with five or six pairs of visible scales. Lateral buds similar but much smaller.

Capsules 5–8 cm across, leathery, light brown, mostly smooth, with one or two

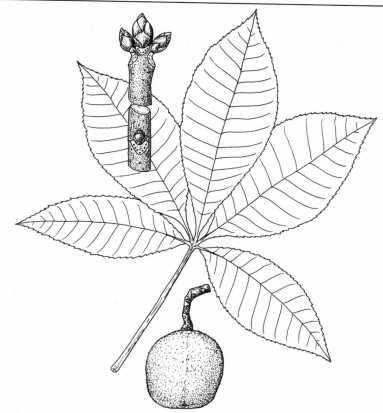

FIGURE 107   Yellow buckeye, *Aesculus octandra*. Sections of twig showing terminal bud, lateral buds, and leaf scars; palmately compound leaf; fruit.

large, lustrous, dark chestnut brown seeds, maturing in late summer or early autumn.

Rich, moist woods; common, mostly at higher elevations, throughout.

2. *Aesculus sylvatica*
painted buckeye

A shrub or small tree, commonly forming thickets.

Leaflets five, narrowly elliptical to lanceolate, 10–15 cm long, finely and often double toothed, pointed at apex, wedge shaped at base, yellowish green above, green and glabrous or somewhat hairy beneath.

Twigs orange green at first, later turning light reddish brown, glabrous.

Terminal bud 6–12 mm long, ovoid, pointed, light reddish brown, with several pairs of overlapping scales.

Capsules light brown, smooth, 2–4 cm across, containing usually a single dark brown seed, maturing in late summer or early autumn.

Riverbanks, low woods; Cherokee and Jackson Cos., NC; SC; GA.

Buckthorn Family (Rhamnaceae)

Trees or shrubs with simple leaves, in our species alternate and deciduous; flowers small and regular, in clusters; fruit a drupe or capsule. A large family, but more common to the west and south of our area; buckthorn, ceanothus.

Rhamnus (buckthorn)

Shrubs or small trees.

Leaves alternate (in our species), deciduous, simple, pinnately veined, toothed or entire, *lustrous green* on upper side.

Twigs slender, angled below nodes.

Buds moderate in size, solitary, sessile (without stalks), *naked* in our species. Terminal bud present, larger than laterals.

Leaf scars crescent shaped to half elliptical. Bundle scars three. Stipule scars very small, or stipules sometimes persistent.

Fruit a small, berrylike drupe containing two or three seedlike nutlets.

1 Leaves very finely or obscurely toothed; lower surface of leaves glabrous; terminal bud 5–8 mm long, light brown or yellowish brown . . . . (1) *R. caroliniana*
1' Leaves entire or wavy margined; lower surface of leaves more or less hairy; terminal bud 3–5 mm long, reddish brown or orange brown . . . (2) *R. frangula*

	1. *Rhamnus caroliniana*
FIGURE 108	Carolina buckthorn

Leaves elliptical to oblong, 6–15 cm long, about one-third as wide, very finely toothed to nearly entire, short pointed or somewhat long pointed at apex, broadly wedge shaped to rounded at base, lustrous green above, paler and almost glabrous beneath; main lateral veins ending in leaf margin but *upcurving near margin.*

Twigs reddish brown to light brown, glabrous or with silky hairs toward tip; vigorous growth often branching in first year.

Terminal bud 5–8 mm long, slender, *naked,* light brown, *woolly.* Lateral buds brown-hairy.

Leaf scars small, moderately raised. Long, slender stipules often persisting.

Fruit a black, globose, berrylike drupe, 5–7 mm across, containing three seedlike nutlets.

Rich woods, shaded hillsides; central mountains, NC; Pickens Co., SC; GA; GSMNP.

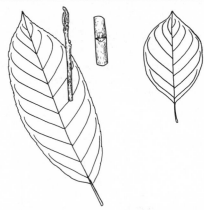

FIGURE 108   *Rhamnus*. Left: Carolina buckthorn, *R. caroliniana*. Leaf, twig with naked
               terminal bud, part of twig with lateral bud and leaf scar. Right: leaf of glossy
               buckthorn, *R. frangula*.

| | 2. *Rhamnus frangula* |
| FIGURE 108 | glossy buckthorn |

Leaves oblong to obovate, 3–7 cm long, a little more than half as wide, *entire or
    slightly wavy* along margin, abruptly short pointed to somewhat rounded at
    apex, broadly wedge shaped to rounded at base, dark green and glabrous above,
    hairy beneath, at least along midrib and larger veins; lateral veins prominent,
    parallel and nearly straight but upcurving near margin.
Twigs reddish brown or brownish red, hairy when young.
Terminal bud 3–5 mm long, naked, densely woolly with reddish brown or orange
    brown hairs. Lateral buds smaller.
Fruits globose, about 6 mm across, red turning dark purple or black, with *two*
    nutlets.
Native of Eurasia, rare escape in Buncombe Co., NC.

| | *Ceanothus americanus* |
| FIGURE 109 | New Jersey tea |

A low shrub, usually with several stems from a large, red rootstalk, the central stem
    woody but the branches often nearly herbaceous.
Leaves alternate, deciduous, simple, ovate, 3–8 cm long, shallowly toothed, with
    three principal veins from base, dull green above, hairy to nearly glabrous
    beneath.

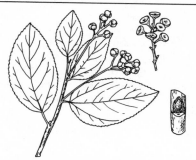

FIGURE 109   New Jersey tea, *Ceanothus americanus*. Twig with leaves and fruit clusters; persistent saucerlike bases after fruits have dropped; lateral bud and leaf scar.

Twigs slender, rounded, yellowish green to tan, dying back in winter.

Buds solitary, small, with about four scales showing, the first two scales almost as long as the bud. Terminal bud present.

Leaf scars small, half elliptical. Bundle scar one. Stipule scars present.

Fruit a dry, three-lobed capsule, splitting lengthwise into three nutlets, in terminal or axillary clusters, falling away at maturity and leaving a *saucerlike base persisting* through winter.

Roadsides, clearings, edges of woods; common, throughout.

Grape Family (Vitaceae)

Mostly woody vines, climbing by means of tendrils; sap watery, acid; leaves alternate, simple, palmately veined or lobed, or compound; flowers produced in clusters opposite the leaves; fruit a berry; grape, ampelopsis, Virginia creeper.

Vitis (grape)

Woody vines, climbing high or trailing, *clinging to support by means of tendrils,* each tendril opposite a leaf or a leaf scar (unless replaced by a fruit scar).

Leaves alternate, deciduous, simple, *palmately veined* and often *lobed;* petioles and tendrils with sour taste.

Stems usually with *loose and shreddy bark.* Pith *brownish,* showing *a woody bar of tissue at each node,* evident when stem is carefully split or cut lengthwise (except in *V. rotundifolia*).

Buds often widely divergent, with two to four scales visible. Terminal bud absent.

Leaf scars half circular to crescent shaped. Bundle scars several, in a C-shaped row, often indistinct. Stipule scars present, long and narrow.

Fruit a juicy berry, in branched clusters opposite the leaves, ripening in late summer or autumn.

1 Tendrils simple, not branched; pith continuous, without a woody partition at each node; older bark close fitting, not shredding . . . . (1) *V. rotundifolia*
1′ Tendrils branched; pith showing a woody bar or partition at each node if stem is split lengthwise; older bark peeling in shreds:
  2 Stems with a tendril or flower cluster or fruit cluster at three or more successive nodes; leaves uniformly covered beneath with a feltlike layer of rusty hairs concealing leaf surface . . . .(2) *V. labrusca*
  2′ Stems lacking tendril or flower cluster or fruit cluster at every third node; leaves glabrous or variously hairy beneath, but not covered with a feltlike layer of rusty hairs, the leaf surface not concealed:
    3 Lower surface of leaves greenish and glabrous or merely short-hairy, at least along veins, the hairs usually erect, not tangled and cobwebby:
      4 Young stems rounded, glabrous; petioles and veins on undersurface of leaves essentially without hairs . . . . (3) *V. vulpina*
      4′ Young stems angled, woolly-hairy; petioles and veins on undersurface of leaves soft-hairy . . . . (4) *V. baileyana*
    3′ Lower surface of leaves bluish white glaucous or cobwebby-hairy when young, the hairs often thin or scattered at maturity but persisting along the veins:

5 Young stems angled, densely short-hairy .... (5) *V. cinerea*
5' Young stems rounded, glabrous or with only a few curly or straight
   hairs:
    6 Leaves beneath rusty brown with loose or tangled hairs
      .... (6) *V. aestivalis*
    6' Leaves beneath whitish or bluish white, often nearly without hairs
      .... (7) *V. aestivalis* var. *argentifolia*

FIGURE 110 | **1. *Vitis rotundifolia***
           | muscadine

Leaves *much smaller than those of other grapes,* 4–8 cm across, nearly circular to
   broadly ovate in general outline, coarsely toothed, short pointed at apex, heart
   shaped at base, *green and glossy on both sides,* but with cobwebby hairs in axils
   of veins beneath.
Stems brown or greenish brown, glabrous or sparingly hairy toward tip; young
   stems angled, older stems striated longitudinally; bark *tight,* not loose and
   shreddy as in other grapes; lenticels abundant and conspicuous, appearing as
   pale dots; *tendrils unbranched.* Pith brown, *without firm partitions at nodes.*
Fruit large, 1–2 cm across, purplish to bronze colored, with thick skin and firm
   flesh; few in a cluster; sweet but skin tough.
Moist open woods, sandy streambanks; occasional, essentially throughout.

FIGURE 110 | **2. *Vitis labrusca***
           | fox grape

Leaves circular to ovate in outline, 10–20 cm long and wide, toothed and usually
   with two short lateral lobes, long pointed at apex, deeply heart shaped at base,
   dull green and glabrous above, lower surface *persistently woolly with a layer of
   rusty or yellowish hairs* concealing the leaf surface.
Young stems densely rusty-hairy; older bark loose and soon shredding. Pith with
   firm partition at each node. Tendrils or flower clusters and fruit clusters *at three
   or more successive nodes;* tendrils hairy.
Fruit large, 12–20 mm across, usually dark purple, occasionally red, sweet
   when ripe.
Woods, roadsides, thickets; fairly common, throughout.

FIGURE 110 | **3. *Vitis vulpina***
           | frost grape

Leaves more or less heart shaped, 5–15 cm long, nearly as wide, coarsely toothed,
   unlobed or shallowly and indistinctly lobed, short pointed or long pointed at
   apex, deeply and narrowly heart shaped at base, glossy green above, *bright green*

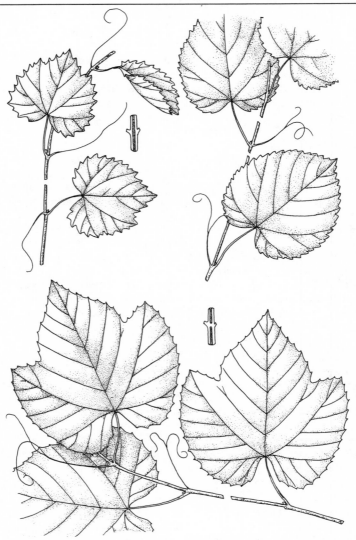

FIGURE 110   *Vitis.* Upper left: muscadine, *V. rotundifolia.* Stem with leaves and
               unbranched tendrils; lengthwise section of stem through a node showing
               continuous pith. Upper right: frost grape, *V. vulpina.* Sections of stem with
               leaves and branched tendrils. Below: fox grape, *V. labrusca.* Leaves, stem,
               and branched tendrils; lengthwise section of stem through node showing
               woody bar of tissue.

beneath, hairy when young, *glabrous or nearly so at maturity* except for tufts of hairs in axils of lower veins.

Stems brown and glabrous; young stems essentially *circular* in cross section, not angled; tendrils branched; *no tendril opposite each third leaf or leaf scar;* older bark slightly shreddy. Pith with partitions, 2–5 mm thick, at nodes.

Fruit *small,* 5–10 mm across, black, mostly without bloom, sweet after frost.

Low woods and thickets; common, throughout.

FIGURE 111	4. *Vitis baileyana* possum grape

Leaves ovate to nearly circular in general outline, 6–10 cm long, rather finely toothed, unlobed or with three short lobes near apex, strongly heart shaped at base, dull green above, paler beneath; *petioles and lower veins densely hairy* with grayish hairs.

Stems *angled* and usually woolly in first year; *tendrils lacking opposite every third leaf or leaf scar;* older bark loose and peeling. Pith with nodal partitions 2–6 mm thick.

Fruit small, 5–8 mm across, black, sweet.

Wooded slopes and streambanks at lower elevations; infrequent, throughout.

FIGURE 111	5. *Vitis cinerea* sweet winter grape

Leaves roundish to broadly ovate in general outline, 10–20 cm long, irregularly toothed, either unlobed or with two small lateral lobes, short pointed or long pointed at apex, narrow, rounded sinus at base, dull dark green above, grayish green beneath; *lower surface covered with ashy gray, cobwebby, and straight hairs* when young, some hairs remaining at maturity, especially on veins; petioles grayish hairy.

Stems angled or *ridged, permanently ashy gray or rusty gray* with layer of short hairs; *a tendril lacking at each third node;* older bark shredding into long, thin, loose strips. Pith with nodal partitions less than 3 mm thick.

Fruit 4–9 mm across, black, with or without bloom, sweet when mature.

Low woods, bottomlands, streambanks; Henderson and Transylvania Cos., NC; Oconee Co., SC; GSMNP.

FIGURE 111	6. *Vitis aestivalis* summer grape

Leaves 8–20 cm long and wide, shallowly or deeply lobed, coarsely toothed; upper surface dull green; lower surface greenish, not strongly whitened but more or less *cobwebby with rusty hairs;* degree of hairiness and hence color of lower surface vary considerably.

FIGURE 111   *Vitis.* Above left: possum grape, *V. baileyana.* Stem with leaves and branched tendrils. Above right: summer grape, *V. aestivalis.* Leaves showing variation in degree of lobing. Below: sweet winter grape, *V. cinerea.* Leaves showing variation in leaf shape.

Stems *circular* in cross section, not angled, marked with fine longitudinal lines, brownish, glabrous or with scattered hairs; *no tendril opposite each third leaf or leaf scar;* older bark loose and shreddy. Pith with firm partitions at nodes, 3–4 mm thick.

Fruit 5–11 mm across, black or purplish black, usually with bloom; edible when ripe, but may be dry and puckery.

Low woods, streambanks, roadsides, thickets; common, throughout.

7. *Vitis aestivalis* var. *argentifolia*
silver-leaf grape

Similar to number 6, but undersurface of leaves strongly whitened and sometimes nearly hairless.

Stems often bluish glaucous, especially when young.

Low woods and streambanks; mountains of NC; also GSMNP.

Ampelopsis (ampelopsis)

High-climbing vines or often shrubby, with or without tendrils opposite the leaves.

Leaves alternate, deciduous, simple or compound.

Stems soft-wooded, angled or nearly circular in cross section, bearing tendrils at some nodes, or nearly without tendrils; bark *close,* not shreddy as in all grapes except one, marked by numerous lenticels; pith *white,* at first continuous, later with *thin plates* extending from outer wall toward center.

Buds small or concealed beneath bark.

Leaf scars circular or nearly so. Bundle scars many, arranged in an ellipse, often indistinct. Stipule scars long and narrow.

Fruit a berry having thin flesh and containing two to four seeds; in drooping clusters; flesh dry, not edible.

1 Leaves compound, leaflets 2–6 cm long . . . . (1) *A. arborea*
1′ Leaves simple, 6–11 cm long. . . . . (2) *A. cordata*

1. *Ampelopsis arborea*
pepper-vine, thunderberry

A high-climbing vine, or often shrubby.

Leaves mostly *twice compound.* Leaflets ovate, 2–6 cm long, coarsely toothed, long pointed at apex, broadly wedge shaped, straight across, or rounded at base, glabrous on upper surface except for hairs along margin, paler beneath and hairy along midrib and larger veins.

Stems nearly circular in cross section, medium brown, glabrous; tendrils few or none.

Berries black, 6 mm across, two or three seeded.

Low woods; Madison Co., NC.

2. *Ampelopsis cordata*
heartleaf ampelopsis

Usually a vine climbing by means of unbranched tendrils, sometimes shrubby and almost without tendrils.

Leaves simple, ovate to somewhat triangular, 6–11 cm long, coarsely toothed with sharply tipped teeth, short pointed to long pointed at apex, *heart shaped* to rounded at base, dark yellowish green above, paler and usually hairy beneath.

Stems moderately stout, flexible, slightly angled in cross section, with ridges running downward from nodes, yellowish green to brownish, glabrous; lenticels prominent; tendrils few, twining.

Berries flattened globe shaped, 8–11 mm across, green turning orange, then pinkish purple, and finally blue when ripe.

Low woods; Madison Co., NC; GSMNP.

FIGURE 112 | *Parthenocissus quinquefolia*
Virginia creeper

A climbing vine, sometimes mistaken for Poison-ivy but distinguishable by its *tendrils* and *palmately compound leaves*.

Leaves alternate, deciduous, long petioled. Leaflets three to seven, usually *five,* elliptical to obovate or oblong, 4–15 cm long, coarsely toothed, mostly above middle, pointed at apex, wedge shaped at base, dull green above, paler beneath.

Tendrils opposite the leaves or leaf scars, absent from every third node; long, slender, branched, some or all *ending in small adhesive disks.*

Stems with bark not shredding or flaking; young stems brown, glabrous or thinly hairy; old stems may have coarse aerial roots. Pith green to white, continuous, without woody partitions at nodes.

Buds commonly two together at a node, conical, blunt, with several scales visible, the outer scales often keeled.

Leaf scars raised, nearly circular. Bundle scars numerous, indistinct. Stipule scars long and narrow.

Fruit a *dark blue berry,* resembling small grape, 5–8 mm across, in clusters; not edible; ripe in July or August.

Dry or moist woods, rocky banks; common, throughout.

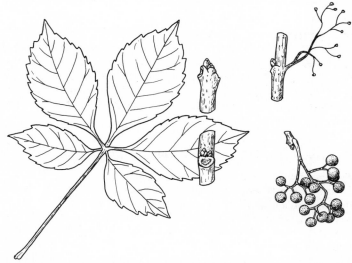

FIGURE 112   Virginia creeper, *Parthenocissus quinquefolia*. Palmately compound leaf; detailed sections of stem showing bud and leaf scar at end of stem, collateral buds and leaf scar; stem with tendrils ending in adhesive disks; and fruit cluster.

Linden Family (Tiliaceae)

Trees or shrubs, often with fibrous bark; leaves alternate in two rows, in our species deciduous, simple, palmately veined, toothed, mostly unequal at base; flowers fragrant, regular, in long-stemmed clusters attached near center of a narrow bract (our species); fruit small, dry, nutlike. Represented in our area by one genus.

Tilia (basswood)

Medium-sized to large trees with tough, fibrous inner bark.

Leaves alternate, two-ranked, deciduous, simple, more or less heart shaped, *palmately veined,* toothed, usually *unsymmetrical at base.*

Twigs slender to moderate, zigzag, circular in cross section, tough and flexible; older twigs showing a ring of triangles in the bark (bast fibers), evident when twig is carefully cut crosswise.

Buds solitary, mostly *red or green,* with two or three scales visible, somewhat ovoid but *lopsided, the first (outer) scale bulging;* buds mucilaginous if chewed; terminal bud absent.

Leaf scars two-ranked, half circular to half elliptical or crescent shaped. Bundle scars three or more, in a single row or scattered. Stipule scars present, large, unequal.

Fruit a *small nutlet,* resembling a pea but having a nutlike hardness, attached to a leafy bract.

The *Checklist of United States Trees (Native and Naturalized)* (1979) accepts three native species in the genus *Tilia.* All three are native to the southern Appalachians and are keyed and described in the following.

1 Lower surface of mature leaves glabrous except for a few simple hairs on
   veins and tufts of hairs in vein axils . . . . (1) *T. americana*
1' Lower surface of mature leaves more or less hairy between veins as well as on
   them, some or all of the hairs branched and star shaped:
     2 Lower surface of leaves pale green to white or brownish, with a dense,
       feltlike coating of hairs . . . . (2) *T. heterophylla*
     2' Lower surface of leaves medium green, thinly to abundantly hairy but not
       with a feltlike coating of hairs . . . . (3) *T. caroliniana*

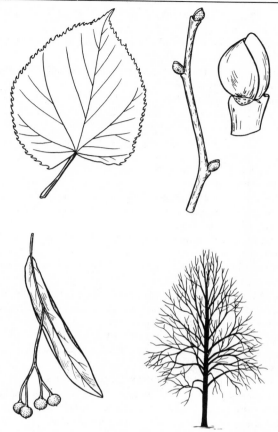

FIGURE 113   American basswood, *Tilia americana*. Above: Leaf, twig, enlarged end of
twig showing topmost lateral bud, leaf scar, and end of season's growth.
Below: fruit attached to a leafy bract, silhouette of mature tree.

FIGURE 113	1. *Tilia americana* American basswood

Bark deeply and closely furrowed, resembling that of white ash (*Fraxinus
americana*).

Leaves ovate to nearly circular, 7–20 cm long, sharply toothed, short pointed or
abruptly long pointed at apex, heart shaped, straight across, or often unequal at

base, dark green and glabrous above, *mostly glabrous beneath* but slightly hairy along veins and with *tufts of hairs in axils of lateral veins.*

Twigs red or reddish brown to yellowish brown, or sometimes reddish above and pink to greenish beneath, glabrous.

Buds 5–7 mm long, blunt or pointed, usually same color as twigs, glabrous, with two or three scales exposed.

Moist woods on bottomlands and upland slopes; occasional, NC.

2. *Tilia heterophylla*
white basswood

Bark thick, deeply furrowed, with grayish brown, scaly ridges.

Leaves broadly ovate to nearly circular, 8–20 cm long, finely toothed, short pointed or long pointed at apex, heart shaped, unequal, or straight across at base; upper surface dark green and glabrous; lower surface *densely hairy with a coating of whitish or brownish hairs.*

Twigs red to green, commonly reddish on side facing light, greenish beneath, *glabrous.*

Buds 5–9 mm long, dark red to green, with two or three scales exposed, the first or lowest scale bulging.

Moist woods on streambanks and upland slopes; fairly common, throughout.

3. *Tilia caroliniana*
Carolina basswood

Bark deeply and closely furrowed.

Leaves ovate to heart shaped, 8–18 cm long, sharply toothed, the teeth fringed with hairs, abruptly long pointed at apex, heart shaped or unequal at base; upper surface dark green and glabrous; lower surface with few to many branched hairs, as well as simple hairs, but *not densely hairy.*

Twigs reddish brown to gray; *young twigs grayish hairy.*

Buds pointed, 5–7 mm long, reddish, usually glabrous but sometimes slightly hairy.

Moist, low or upland woods; central mountains of NC.

Mallow Family (Malvaceae)

A large family of herbs, shrubs, and trees, only one woody member found in our area; leaves alternate, deciduous, simple, mostly palmately veined; flowers large, showy; fruit a capsule.

FIGURE 114	*Hibiscus syriacus* rose-of-Sharon, shrubby althea

A shrub or small tree.

Leaves alternate, deciduous, simple, 5–12 cm long, usually *palmately three-lobed,* with three principal veins from the base, coarsely toothed, the teeth rounded or somewhat pointed; broadly wedge shaped at base, medium green above, a little paler beneath.

Twigs slender, gray to grayish green, short-hairy or glabrous, with ridges extending downward from nodes, *enlarged and fluted at tip;* often bearing large circular scars left by fallen flowers.

Buds not clearly evident, *partly or wholly concealed in tissue above leaf scar,* some apparently borne in clusters at ends of twigs. Terminal bud present.

Leaf scars raised, half circular to nearly circular, usually several crowded toward twig tip. Bundle scars about four, indistinct. Stipules long, slender, threadlike, persistent or leaving small raised scars.

Fruit a large, ovoid capsule, 2–3 cm long, splitting open when ripe and releasing seeds; usually present and conspicuous in winter.

Native of eastern Asia, persistent after cultivation and occasionally escaping to roadsides and thickets.

Tea Family (Theaceae)

Shrubs or trees, chiefly in the tropics, with alternate, simple, toothed leaves, in our species deciduous; flowers solitary, large, showy; fruit a woody capsule. Represented in our area by one genus.

Stewartia (stewartia)

Shrubs or small trees with shreddy bark.

Leaves alternate, deciduous, simple, finely or obscurely toothed.

Twigs slender, circular in cross section, with *spongy* pith.

Buds spindle shaped, flattened, more or less silky, with two outer scales. Terminal bud usually present. Lateral buds sometimes superposed.

FIGURE 114   Rose-of-Sharon, *Hibiscus syriacus*. Leaves, twig, and fruit; detailed sections
                  of twig showing leaf scar, partially submerged buds, and persistent
                  threadlike stipules.

Leaf scars half circular or crescent shaped, open side up. Bundle scar *one, large, raised,* at top of leaf scar. Stipule scars lacking.

Fruit a woody *capsule,* five-celled, separating into five parts, each part containing one to four seeds; maturing by autumn.

1 Buds 4 mm long or more, densely hairy, the lateral buds nearly or entirely
   concealed by enfolding wings on leaf petiole; leaves 6–15 cm long
   ....(1) *S. ovata*
1′ Buds less than 3 mm long, glabrous or only slightly hairy, not at all concealed by narrow wings on leaf petiole; leaves mostly 4–8 cm long
   ....(2) *S. malacodendron*

	1. *Stewartia ovata*
FIGURE 115	mountain camelia, mountain stewartia

A shrub or small tree. Bark furrowed, grayish brown.
Leaves ovate to broadly elliptical, 6–15 cm long, finely or obscurely sharp toothed, at least above middle, long pointed at apex, broadly wedge shaped to rounded

FIGURE 115   Mountain camelia, *Stewartia ovata*. Leaf; cilia along leaf margin; section of twig showing leaf scar and flattened and twisted terminal and lateral buds.

at base, dark green and mostly glabrous above, lighter green and hairy beneath; margin of leaf petioles with *upturned wings,* these nearly to entirely concealing the lateral buds.

Twigs light brown to reddish brown, frequently tinged with purple, glabrous or sparingly hairy; bark often shreddy; *pith spongy.*

Terminal bud 4–10 mm long, *silvery-silky, flattened and twisted,* with two outer scales, but the scales obscured by silky hairs. Lateral buds similar but smaller, the scales not readily apparent.

Capsules *strongly five-angled,* pointed, about 2 cm long, containing several angled or winged seeds.

Rich, wooded ravines and streambanks; southwest mountains of NC; SC; GA; GSMNP.

2. *Stewartia malacodendron*
silky camelia

Usually a shrub. Bark smooth, dark brown, with lighter colored patches underneath, appearing *mottled.*

Leaves elliptical to elliptical-oblong, mostly 4–8 cm long, obscurely toothed and ciliate, short pointed at apex, wedge shaped at base, bright green and glabrous above, paler and softly hairy beneath; petioles somewhat winged, but not with upturned wings as in *S. ovata.*

Twigs light brown or dark brown to gray, often finely hairy.

Buds *small,* mostly less than 3 mm long, slender, glabrous or slightly hairy.

Capsules faintly angled, 10–15 mm long, containing several shiny seeds, not angled or winged.

Low-lying woods and streambanks; Avery and Macon Cos., NC.

St. Johnswort Family (Hypericaceae)

Herbs or shrubs with opposite, simple, deciduous, entire leaves bearing many dotlike translucent glands; flowers small or large, mostly yellow, solitary or in clusters; fruit a capsule containing many seeds. Represented in our area by one genus.

Hypericum (St. Johnswort)

Erect shrubs with shreddy bark (most of our species herbaceous).

Leaves *opposite*, simple, *dotted with translucent glands,* nearly without petioles, often with *tufts of smaller leaves in axils.*

Twigs slender, more or less *two-angled and flattened,* especially just below nodes.

Buds solitary, very small, narrowly conical, with two or more scales visible.

Leaf scars opposite, small, somewhat raised, triangular. Bundle scar one. Stipule scars absent, but opposing leaf scars connected by a transverse ridge.

Fruit a dry capsule, splitting lengthwise, containing many small seeds; in mainly terminal clusters, persisting into winter.

1 Leaves generally linear, the larger ones 3–7 mm wide; capsules less than
  6 mm long . . . . (1) *H. densiflorum*
1' Leaves generally oblong, the larger ones 7–14 mm wide; capsules more than
  6 mm long . . . . (2) *H. prolificum*

1. *Hypericum densiflorum*
dense-flowered St. Johnswort

Leaves linear to elliptical, largest 2–5 cm long, usually *less than 7 mm wide,* entire, the margin slightly rolled under, pointed at apex, wedge shaped at base, light green above, paler beneath, glabrous on both sides; *upper surface with scattered glandular dots.*

Twigs slender, two-edged, brown, glabrous.

Buds very small, green, glabrous, usually developing into clusters of leafy stems in axils of main leaves.

Capsules three-celled, narrowly conical, 3–6 mm long, brown, in dense, flat-topped clusters.

Wet or dry woods and balds; infrequent, throughout.

2. *Hypericum prolificum*
shrubby St. Johnswort

Leaves elliptical, narrowly oblong, or oblanceolate, the largest 3–5 cm long, *usually more than 7 mm wide*, entire, the margin slightly rolled under, blunt or pointed at apex, wedge shaped at base, pale green above, lighter green beneath, glabrous on both surfaces; upper side dotted with *small, scattered translucent glands.*

Twigs usually branched at or near base, slender, *two-edged,* light ashy brown to yellowish brown or reddish brown, glabrous; pith brown.

Buds small, green, glabrous; lateral buds usually developing into short leafy branches, appearing as clusters of small leaves in axils of main leaves.

Capsules three-celled, cylindrical to narrowly ovoid, 6–13 mm long, brown, in narrow open clusters.

Streambanks, rocky slopes, edges of moist or dry woods; fairly common, throughout.

Mezereum Family (Thymelaeaceae)

Trees or shrubs with very tough, acrid inner bark; leaves alternate, deciduous, simple, entire; flowers generally small, inconspicuous; fruit a berrylike drupe. Represented in our area by one species.

	Dirca palustris leatherwood

A widely branching shrub with *tough bark.*

Leaves alternate, deciduous, simple, broadly elliptical to obovate, 4–9 cm long, entire, short pointed at apex, wedge shaped to rounded at base, green and glabrous on both sides; *petioles enlarged at base, concealing the buds.*

Twigs slender, rounded, enlarged at nodes, appearing *jointed,* light brown to olive green, with silky hairs; bark of twigs pliant but strong, almost impossible to break by hand; wood soft and weak.

Buds solitary, short, conical, brown, velvety, with about four scales visible. Terminal bud absent.

Leaf scars *deeply U-shaped,* extending nearly around buds. Bundle scars five.

Fruit a green to red ellipsoid drupe, 6–12 mm long, ripening in June or July and soon falling.

Rich woods; Ashe, Clay, McDowell, and Madison Cos., NC.

Oleaster Family (Elaeagnaceae)

Shrubs or trees, mostly silvery-scaly; twigs often thorny; leaves alternate (in our species) or opposite, deciduous, simple, entire; flowers small, solitary or clustered; fruit drupelike. Represented in our area by one species.

FIGURE 116	*Elaeagnus umbellata* autumn-olive, autumn elaeagnus

A large shrub, often with *thorny branches.*

Leaves alternate, deciduous, simple, broadly elliptical to ovate or ovate-lanceolate, 3–8 cm long, entire or wavy margined, long pointed at apex, wedge shaped or rounded at base, green above, *uniformly silvery beneath.*

Twigs slender, circular in cross section, *silvery, or silvery and golden brown,* with a

FIGURE 116   Autumn-olive, *Elaeagnus umbellata*. Leaf; detailed sections of twig showing terminal bud, lateral bud, and sharp-pointed spur twig.

complete covering of minute scaly particles; spur twigs common, some sharply pointed.

Terminal bud present, narrow, brownish or silvery, indistinctly scaly. Lateral buds sometimes two together (collateral).

Leaf scars small, raised, half circular to nearly circular. Bundle scar *one*. Stipule scars absent.

Fruit berrylike, nearly globose, 6–9 mm across, silvery or silvery brown, finally *red*.

Native of eastern Asia, occasionally escaped from cultivation to moist woods and roadsides.

Ginseng Family (Araliaceae)

Mostly woody plants, also a few herbs; leaves usually alternate, deciduous or evergreen, simple or pinnately compound; stems frequently prickly or spiny; flowers small, usually numerous, greenish or whitish, in clusters; fruit a purple or black drupe with thin flesh; English ivy, devils-walkingstick.

FIGURE 117	*Hedera helix* English ivy

A climbing or trailing vine with *aerial rootlets*.

Leaves alternate, simple, *evergreen,* thick and leathery, palmately veined, 5–10 cm

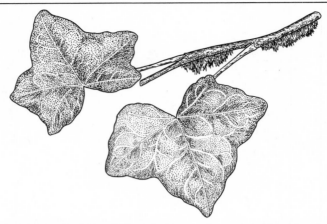

FIGURE 117   English ivy, *Hedera helix*. Evergreen leaves, stem with aerial rootlets.

long, ovate to circular in outline, *variously three- to five-lobed or angled* (leaves of flowering stems tend to be without lobes), blunt to short pointed at apex, heart shaped to wedge shaped at base, dark green and with pale green or whitish veins above, slightly paler and glabrous beneath.

Stems moderately slender, greenish, covered with branched or star-shaped hairs when young, later nearly glabrous; pith greenish, spongy.

Buds conical, somewhat flattened, 1–4 mm long, brown-hairy.

Leaf scars slightly raised, U-shaped, extending more than halfway around stem. Bundle scars five to seven. Stipule scars absent.

Fruit a globose, nearly black berry, 7–8 mm across, containing three to five seeds, but fruits produced only on adult form of plant, and this form rarely seen.

Native of Europe, often cultivated and occasionally escaping into rich woods and waste places.

FIGURE 118 | *Aralia spinosa*
| devils-walkingstick, Hercules club

A large, erect shrub or (occasionally) small tree with *prickly stems and leaf petioles.*

Leaves alternate, deciduous, two or three times compound, *very large;* leaflets many, 4–10 cm long, elliptical to ovate, toothed, short pointed or long pointed at apex, rounded to somewhat heart shaped at base; base of petiole sheathing stem.

Twigs and branches *stout,* set with strong, spiny prickles scattered irregularly over bark or, especially on older stems, borne mostly in a crownlike row at or just below the nodes; prickles strong, straight or curved.

FIGURE 118   Devils-walkingstick, *Aralia spinosa*. Left: Sections of stem with terminal
bud, lateral buds, leaf scars more than half encircling stem, and stout
prickles. Right: portion of twice-compound leaf.

Terminal bud large, conical, bluntly pointed, few scaled. Lateral buds smaller, set a
    short distance above leaf scar.
Leaf scars *unusually large,* V-shaped or U-shaped, extending halfway or more
    around twig. Bundle scars many, prominent, in a single row. Stipule scars absent.
Fruit a globose or ovoid black berrylike drupe, 4–6 mm long, in large clusters.
Moist woods and edges of woods; occasional, throughout.

Tupelo-Gum Family (Nyssaceae)

Trees with alternate, simple, deciduous, entire or few-toothed leaves;
flowers small, inconspicuous; fruit thin fleshed, drupelike, with ribbed
seeds. Represented in our area by one species.

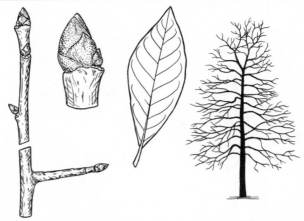

FIGURE 119   Black tupelo, *Nyssa sylvatica*. Twig showing wide-angled branching, enlarged terminal bud and uppermost lateral bud, leaf, and silhouette of mature tree.

| | *Nyssa sylvatica* |
| FIGURE 119 | black tupelo, blackgum |

A tree with an erect trunk, many of the branches horizontal or the lower ones drooping. Mature bark *very rough,* grayish brown to blackish, deeply furrowed, *the ridges broken into oblong blocks.*

Leaves alternate, deciduous, simple, narrowly obovate to broadly elliptical, 5–15 cm long, entire, or mostly entire but occasionally with one or two large teeth, usually *broadest above middle,* abruptly long pointed or short pointed at apex, wedge shaped at base, dark green and glossy above; foliage turning brilliant colors in autumn.

Twigs slender to moderately stout, reddish brown to reddish green, glabrous or slightly hairy; *short spur twigs common.* Pith solid but *chambered* (sections of whitish tissue alternating with firmer cross bands), best seen under magnifying lens.

Terminal bud present, somewhat larger than laterals, 4–8 mm long, ovoid, reddish brown, with three to six scales exposed, the outer scales *glabrous and shiny,* the inner scales short-hairy. Lateral buds solitary or superposed, stalkless or short stalked, usually set at wide angle to twig, often lacking on short growths.

Leaf scars broad and rounded, half circular to heart shaped. Bundle scars *three, distinct, lighter colored than leaf scar,* usually sunken below surface of leaf scar. Stipule scars absent.

Fruit an ovoid, dark blue, one-seeded drupe, borne singly or in twos or threes, flesh sour and bitter; ripening in late summer or autumn and soon dropping.

Low woods, upland woods; common, throughout.

Dogwood Family (Cornaceae)

Mostly shrubs or small trees; leaves simple, deciduous, entire, usually opposite; flowers small, white or greenish, in terminal clusters or in dense heads surrounded by showy bracts; fruit a drupe. Only two genera in North America, one represented in our area.

Cornus (dogwood)

Shrubs or small trees.

Leaves opposite (alternate or almost opposite in *C. alternifolia*), deciduous, simple, entire, with *lateral veins strongly and evenly curving toward apex of blade.*

Twigs slender, often highly colored.

Buds with mostly two visible scales (flower buds covered by two opposite pairs of scales in *C. florida*). Terminal bud present, much larger than laterals, or the lateral buds concealed behind raised leaf scars.

Leaf scars narrow, raised, broadly U-shaped, crescent shaped, or half elliptical, with three bundle scars. Stipule scars absent, but opposite leaf scars joined around twig by a descending ridge or flap.

Fruit a fleshy, distinctively colored drupe, containing single bony stone; borne in open, spreading clusters or in dense, headlike clusters.

In winter the dogwoods are similar to some of the viburnums, but leaf scars on young twigs of dogwoods are much raised above twig surface, whereas as in viburnums the leaf scars are only slightly raised.

1 Leaves and leaf scars alternate but mostly crowded toward twig tips
.... (1) *C. alternifolia*
1′ Leaves and leaf scars opposite, rarely crowded toward twig tips:
  2 Pith of two-year-old twigs white; lateral buds concealed behind base of leaf petioles; flower buds often present, terminal, large, somewhat button shaped, on long stalks..................................... (2) *C. florida*
  2′ Pith of two-year-old twigs tan or brown; lateral buds small, triangular, closely appressed, densely brown-hairy; flower buds, if present, not as above .... (3) *C. amomum*

FIGURE 120 | 1. *Cornus alternifolia*
alternate-leaf dogwood

An erect shrub or small tree with widely spreading branches, commonly arranged in irregular horizontal layers.

Leaves *alternate* but usually clustered near twig tip, ovate or obovate, 6–12 cm

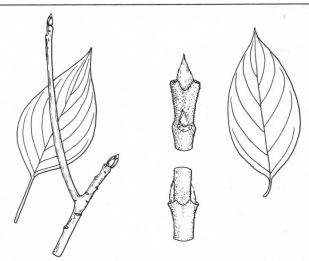

FIGURE 120   *Cornus.* Left: alternate-leaf dogwood, *C. alternifolia.* Leaf, twig showing
vigorous first-year branching. Right: silky dogwood, *C. amomum.* Detailed
sections of stem showing terminal bud, lateral buds, and raised leaf scars;
leaf.

long, entire but often wavy margined, short pointed or long pointed at apex,
wedge shaped at base, green above, pale and covered with short, white appressed
hairs beneath; *the lower veins running nearly parallel to margin, the upper veins
meeting in tip of leaf;* petioles, or most, 2–5 cm long.

Twigs reddish purple or reddish brown to green, glabrous, with unpleasant odor if
cut or bruised; *terminal twig commonly exceeded in length by the lateral twig
just below it;* pith white.

Terminal buds present, all about alike, narrowly ovoid, 4–6 mm long, pointed, reddish purple to greenish, with two to four loosely fitting scales showing; bud
scales keeled, slender pointed. Lateral buds much smaller, often nearly concealed
by persistent base of leaf petioles.

Leaf scars alternate, tending to be crowded toward tip of twig.

Fruit in clusters, blue black, covered with bloom, 6–8 mm across.

Streambanks, wooded slopes, high mountains; common, throughout.

	2. *Cornus florida*
FIGURE 121	flowering dogwood

A small tree with branches widely spreading from a short, often leaning trunk. Bark
thin but rough, made up of *small, checkerlike scales* fitted closely together, reddish brown to nearly black.

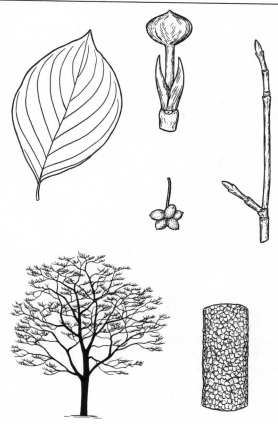

FIGURE 121   Flowering dogwood, *Cornus florida.* Leaf showing arcuate venation; enlarged flower bud with bracts at base of stalk; fruit cluster; twig with terminal leaf buds; silhouette of mature tree; section of trunk showing checkered bark.

Leaves opposite, widely elliptical, mostly 6–12 cm long, entire or slightly wavy margined, short pointed or long pointed at apex, rounded or broadly wedge shaped at base, dark green and very short hairy above, pale and glabrous or finely hairy beneath; lateral veins curving forward, nearly parallel to leaf margin, the uppermost veins meeting in apex of blade; petioles short, 6–18 mm long, their base clasping the stem and concealing the lateral buds.

Twigs green, *purplish,* or reddish, usually glaucous, appearing diamond shaped in cross section.

Terminal buds of two distinct kinds: flower buds *long stalked,* flattened globe shaped, grayish, covered with two pairs of equal-sized scales, the outer pair clearly evident, the inner pair nearly hidden; leaf buds oblong, flattened, pointed, purplish red, with two valvate scales. Lateral buds *concealed behind raised leaf scars.*

Leaf scars opposite, raised above twig surface, borne on base of old petioles persisting in winter.

Flowers in small, dense clusters *surrounded by four white bracts,* each 3–5 cm long, often mistaken for petals.

Fruit in *tight clusters,* ellipsoid, about 1 cm long, *bright red,* shiny, fleshy, occasionally persisting into winter.

Both moist and dry woods; common, throughout.

	3. *Cornus amomum*
FIGURE 120	silky dogwood

A loosely branched shrub.

Leaves opposite, ovate-lanceolate to broadly elliptical, 6–12 cm long, entire, abruptly short pointed to long pointed at apex, rounded or broadly wedge shaped at base, green and glabrous above, *pubescent beneath, the hairs often rust colored.*

Twigs dull purplish or reddish, *silky with whitish or rusty hairs;* pith of two-year-old twigs tan or brown.

Terminal bud conical, 3–5 mm long, light brown, grayish hairy. Lateral buds *clearly visible* on mature growth, 2–3 mm long, sessile, flattened-triangular, closely appressed, densely hairy.

Fruit *light blue,* 5–8 mm across.

Moist or wet woods and streambanks; infrequent, throughout.

White-alder Family (Clethraceae)

Shrubs or trees with alternate, deciduous, simple leaves; stems covered with tufted hairs; flowers terminal, in long, narrow, spikelike fragrant clusters; fruit a capsule containing numerous seeds. A family with only a single genus, its members widely distributed in the world.

FIGURE 122	*Clethra acuminata* mountain pepperbush, cinnamon clethra

A large shrub or occasionally a small tree. Bark on older stems *splitting and peeling in thin plates.*

Leaves alternate but mostly crowded toward twig tip, deciduous, simple, thin, elliptical to elliptical lanceolate, 8–20 cm long, finely and sharply toothed, long pointed at apex, wedge shaped or rounded at base, bright green and glabrous above, pale green and slightly hairy or glabrous beneath.

Twigs with a single prominent ridge downward from nodes, light brown or light reddish brown, *densely coated with star-shaped clusters of hairs; commonly producing side growth in first year,* the lateral twigs exceeding the terminal twig in length (plants growing in deep shade may not branch in first year).

Terminal bud present, much larger than laterals, 3–6 mm long, *yellowish silky, appearing naked* but actually having three narrow outer scales about as long as the bud, often shed early. Lateral buds very small, silky-hairy, tending to be crowded toward twig tip, or frequently developing into leafy twigs in first year.

Leaf scars slightly raised, shield shaped to triangular. Bundle scar *one,* large, often protruding above surface of leaf scar. Stipule scars absent.

Fruit an ovoid hairy capsule, about 5 mm long, splitting in three parts, containing many seeds; borne in upright clusters.

Rich woods; common, throughout.

Heath Family (Ericaceae)

A large family, mostly trees or shrubs, but a few members herbaceous; leaves evergreen or deciduous, simple, alternate, opposite, or whorled; flowers often bell shaped or funnel shaped, solitary or in clusters; fruit a capsule or berry. All members of the family grow in acid soil; rhododendron, mountain-laurel, and many others.

FIGURE 122   Mountain pepperbush, *Clethra acuminata*. Twig showing first-year
branching; typical leaf.

FIGURE 123 | *Chimaphila maculata*
pipsissewa, spotted wintergreen

A nearly herbaceous undershrub with erect stems, usually not more than 15–20 cm
high, arising from creeping underground stems.

Leaves opposite, whorled, or scattered, mostly crowded near tip of stem, *evergreen,*
somewhat leathery, lanceolate to narrowly ovate, 2–6 cm long, remotely
toothed, long pointed at apex, rounded to broadly wedge shaped at base, green
striped with greenish white along midrib and often along larger lateral veins.

Fruit a flattened-globose capsule, 4–7 mm across, borne in groups of two to five at
top of terminal stalk above leafy stem.

Dry woods, especially in acid, sandy soils; common, throughout.

Rhododendron (rhododendron, azalea)

Shrubs or small trees; bark usually close.

Leaves alternate but mostly *crowded toward tip of twigs,* evergreen or deciduous, simple, entire or minutely toothed and ciliate.

Twigs slender to stout, circular in cross section, commonly branching from topmost lateral buds clustered around terminal bud, causing branches to radiate out in
false whorls.

FIGURE 123   Pipsissewa, *Chimaphila maculata*. Evergreen leaves, capsules at top of
terminal stalk.

Flower buds and leaf buds separate. Flower buds terminal (but not always present), *much larger than leaf buds* and covered by more scales, usually in a cluster
with several leaf buds. Uppermost lateral buds crowded near twig tip, the lower
laterals smaller or not much developed.

Leaf scars low, not or only slightly raised, triangular or shield shaped, mostly
grouped toward end of current year's growth. Bundle scar one. Stipule scars absent.

Fruit a five-celled capsule, often persisting for a while.

1 Leaves thick and leathery, evergreen:
  2 Leaves abundantly dotted beneath with small, yellowish to brownish
    scales or pits (visible under hand lens)......................(3) *R. minus*
  2' Leaves not dotted beneath with small scales or pits:
    3 Leaves 12–30 cm long, wedge shaped at base; flower buds with long,
      thin leafy bracts.....................................(1) *R. maximum*
    3' Leaves 8–15 cm long, rounded at base; flower buds without leafy
      bracts...(2) *R. catawbiense*
1' Leaves not thick and leathery, deciduous:
  4 Upper surface of leaves bright green or dull green, not lustrous:
    5 Leaves densely grayish-hairy beneath, at least along midrib and larger
      veins, with a mixture of long coarse hairs and short curly hairs
    ....(4) *R. calendulaceum*
    5' Leaves nearly glabrous beneath or merely with coarse or stiff appressed
      hairs on lower midrib:

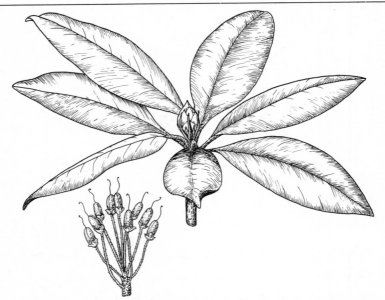

FIGURE 124   Rosebay rhododendron, *Rhododendron maximum*. Twig bearing terminal
flower bud and clustered leaves; terminal cluster of capsules.

    6 Leaves narrowly to broadly elliptical; apex of leaves gradually taper-
       ing to a long point . . . .(5) *R. vaseyi*
    6′ Leaves, or some of them, obovate; apex of leaves tapering to a short
       point or somewhat blunt. . . . . (6) *R. nudiflorum*
 4′ Upper surface of leaves dark green and lustrous:
    7 Young twigs and lower surface of leaves essentially glabrous
       . . . . (8) *R. arborescens*
    7′ Young twigs more or less bristly, at least near tip, with coarse brown
       hairs; lower surface of leaves with long stiff hairs on midrib
       . . . .(7) *R. viscosum*

FIGURE 124	1. *Rhododendron maximum* rosebay rhododendron

A thicket-forming shrub or small tree.
Leaves *evergreen, thick and leathery,* oblanceolate to narrowly elliptical, 12–30 cm
    long, entire, the edges often rolled under, tightly rolled into a cylinder in drought
    or extreme cold; short pointed at apex, *wedge shaped at base,* dark green and
    lustrous above, paler green to brownish and finely hairy beneath.

Twigs stout, dark green and scurfy when young, becoming red and nearly glabrous at maturity.

Flower buds terminal, very large, 25–35 mm long, yellowish green, with many overlapping scales and with *several narrow leafy bracts.* Leaf buds smaller than flower buds, usually lateral, sometimes terminal, solitary, greenish, many scaled.

Flowers white to pink, in large terminal clusters.

Capsules narrowly ovoid, 9–12 mm long, dark reddish brown, with gland-tipped hairs; maturing in September or October and persisting through winter.

Moist or wet woods, along shaded streams, often in dense thickets; our most common rhododendron, throughout.

FIGURE 125 | ## 2. *Rhododendron catawbiense*
Catawba rhododendron

A large shrub or (occasionally) a small tree, often forming thickets.

Leaves *evergreen, thick and leathery,* narrowly to widely elliptical, 8–15 cm long, entire, the edges sometimes rolled under; drooping in winter; rounded to blunt at apex, *rounded at base,* dark green and lustrous above, glaucous or *whitish* and glabrous beneath.

Twigs stout, light green to dark green, becoming red to golden brown in second year.

Flower buds 12–20 mm long, ovoid, dark green to brown, *without leafy bracts;* bud scales edged with hairs and more or less scurfy-hairy over surface. Leaf buds smaller, usually lateral.

Flowers large, pink to bluish purple, in upright clusters.

Capsules linear-oblong, 10–18 mm long, densely covered with reddish brown hairs.

Upland woods, high ridges, heath balds; fairly common, mostly above 3,000 feet, throughout.

FIGURE 125 | ## 3. *Rhododendron minus*
Piedmont rhododendron

An erect shrub with *leaves, petioles, twigs, and buds dotted with many small scales or pits.*

Leaves leathery, persistent, elliptical, 4–12 cm long, entire, the edges usually flat, not revolute, sharply pointed at apex, narrowly to broadly wedge shaped at base, dark green and glabrous above, much paler beneath; glandular dotted on lower side.

Flower buds terminal, 7–10 mm long, ovoid, greenish brown, with many scales exposed. Terminal leaf buds smaller and narrower than flower buds, with a few greenish or brownish scales. Lateral buds much reduced in size or lacking.

Flowers pink to white, often spotted greenish, in terminal clusters.

Capsules 8–10 mm long.

Streambanks, wooded slopes, high ridges; occasional, throughout.

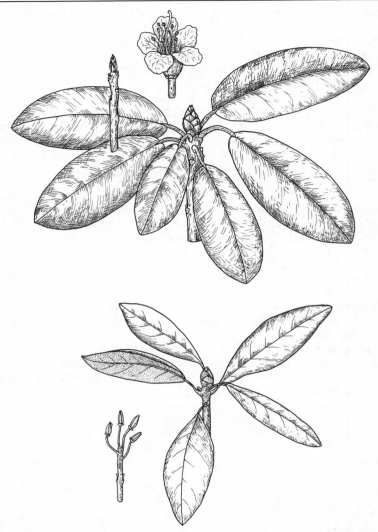

FIGURE 125　　*Rhododendron.* Above: Catawba rhododendron, *R. catawbiense.* Twig bearing terminal flower bud and clustered leaves; section of twig with leaves removed showing terminal leaf bud; typical flower. Below: Piedmont rhododendron, *R. minus.* End of twig with terminal flower bud and clustered leaves; fruit cluster.

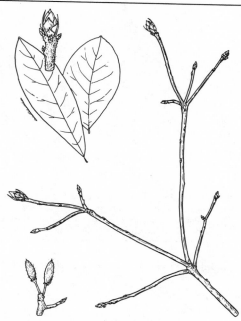

FIGURE 126   Flame azalea, *Rhododendron calendulaceum*. Above: typical leaves, detailed end of twig showing terminal flower bud surrounded by leaf buds. Below: typical fruit. Right: characteristic form of branching in all rhododendrons, the twigs radiating in false whorls.

FIGURE 126	**4. *Rhododendron calendulaceum*** flame azalea

An erect shrub.

Leaves deciduous, oblanceolate to elliptical or narrowly obovate, 5–10 cm long, entire but ciliate on margin, pointed at apex, the tip of midrib protruding, rounded, knoblike; wedge shaped at base, bright to dull green above, pale green and *whitish-hairy beneath* and with appressed bristlelike hairs on lower veins.

Twigs slender, reddish brown to grayish brown, *hairy* with mixture of long and short hairs.

Buds brown, pale green, or reddish green to brownish green. Flower buds ovoid, 8–13 mm long, the scales mostly glabrous except for ciliate margin; lower scales keeled and awned, the lowermost scale often with bristly hairs and long awn. Leaf buds 4–5 mm long, glabrous or nearly so.

FIGURE 127   *Rhododendron.* Left to right: leaves of pinkshell azalea, *R. vaseyi,* pink azalea, *R. nudiflorum,* and swamp azalea, *R. viscosum.* Extreme right: smooth azalea, *R. arborescens.* Twig with terminal cluster of leaf buds; twig with terminal flower bud surrounded by leaf buds; typical leaf.

Flowers *yellow, orange, or red,* appearing early, before leaves are fully grown; not fragrant.

Capsules linear-oblong, erect, 15–30 mm long, more or less hairy but no hairs gland tipped.

Dry or moist, open woods; common, throughout.

FIGURE 127	5. *Rhododendron vaseyi* pinkshell azalea

A tall shrub.

Leaves deciduous, elliptical to elliptical-lanceolate, 5–12 cm long, entire or minutely toothed, with or without cilia, *long tapering* to a pointed apex, wedge shaped at base, dull green and glabrous above or minutely hairy along upper midrib, paler and glabrous beneath except for scattered hairs on lower midrib or sometimes hairs lacking.

Twigs light reddish brown, minutely short-hairy and with a few long, straight hairs when young, later becoming glabrous or nearly so.

Flower buds broadly ovoid, blunt at tip; bud scales hairy along margin but otherwise glabrous, *awnless* or the lower scales short awned.

Flowers deep pink, appearing before the leaves unfold.

Capsules narrowly oblong, 12–24 mm long, *glabrous* or with a few gland-tipped hairs.

Bogs and spruce forests at high elevations, NC.

| FIGURE 127 | 6. *Rhododendron nudiflorum*
pink azalea |

An erect shrub.

Leaves deciduous, obovate to elliptical or oblong, 3–8 cm long, entire but margin ciliate, short pointed at apex, narrowly wedge shaped at base, *dull green* and glabrous on both sides except for *stiff appressed hairs along lower midrib.*

Twigs slender, orange brown to grayish brown, coarsely hairy or nearly without hairs.

Buds pale green, often tinged with red, the scales ciliate but otherwise usually *glabrous,* sometimes short-silky. Flower buds 9–12 mm long, broadly ovoid, the lower scales long awned, the upper ones short awned. Lateral buds 2–4 mm long.

Flowers pale pink to nearly white, appearing before or with unfolding of the leaves; not or only slightly fragrant.

Capsules narrowly oblong, 10–20 mm long, erect, with stiff, appressed hairs, not glandular.

In moist or dry open woods and along streams, mostly at lower elevations, throughout.

| FIGURE 127 | 7. *Rhododendron viscosum*
swamp azalea |

An erect shrub.

Leaves deciduous, obovate to oblanceolate, 2–6 cm long, 7–25 mm wide, entire but ciliate, short pointed or blunt at apex, wedge shaped at base, *dark green and lustrous above,* lighter green and glaucous beneath, *bristly-hairy on lower midrib.*

Twigs slender, dull brown, bristly, especially toward tip.

Flower buds broadly ovoid, 6–10 mm long, reddish, brown, or greenish, minutely *silky;* the upper scales round tipped, the lower bristly and short awned or merely mucronate. Leaf buds small, ovoid, pointed, with several scales.

Flowers white or sometimes pinkish, clammy, appearing after leaves have unfolded; very fragrant.

Capsules oblong-ovoid, 10–20 mm long, with bristly hairs, some hairs glandular.

Low woods, bogs, streambanks; occasional, throughout.

| FIGURE 127 | 8. *Rhododendron arborescens*
smooth azalea |

A tall shrub.

Leaves deciduous, obovate (widest above middle), 3–8 cm long, relatively wider than leaves of other native azaleas, entire, ciliate, short pointed or rather blunt

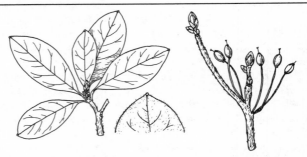

FIGURE 128   Minnie-bush, *Menziesia pilosa*. Twig with clustered buds and leaves; enlarged apex of leaf showing knoblike extension of midrib; twig with clustered buds, chaffy bark, and capsules held erect on slender stalks.

at apex, wedge shaped at base, usually dark green and lustrous above, pale green or whitened beneath, glabrous on both sides except for short hairs on upper midrib and sometimes with a few hairs along petiole and midrib on lower side.

Twigs slender; young twigs yellowish brown, older twigs grayish, *glabrous.*

Buds yellowish green to yellowish brown, often tinged with red or pink. Flower buds narrowly ovoid, 8–13 mm long, the surface of scales *glabrous* but margin ciliate; lower scales long awned.

Flowers white or pinkish, appearing after the leaves are fully grown, very fragrant.

Capsules oblong, 12–16 mm long, densely set with fine glandular hairs.

Mostly along streams and in wet woods; occasional, throughout.

FIGURE 128	*Menziesia pilosa* minnie-bush

A freely branching shrub with shreddy bark.

Leaves alternate, deciduous, simple, mostly crowded near tips of twigs, elliptical to oblanceolate, 3–8 cm long, entire or very minutely toothed, hairy along margin, pointed at apex, *the blade ending in small whitish or reddish, nipple-shaped knob* (mucro); wedge shaped at base, dark green above, pale green or bluish green beneath, with *small nodelike swellings fringed with hairs on lower midrib,* usually visible without magnifying lens.

Twigs slender, stiff, light reddish brown to golden brown, finely hairy and also with scattered long hairs; older bark usually loose and flaky; new twigs from top lateral buds, often appearing whorled.

Buds, or most of them, *borne at or near twig tips.* Flower and leaf buds separate, the larger flower buds with five or six overlapping scales, at least the lower scales strongly keeled, with long-hairy margins and often with long stiff hairs along keel; the smaller leaf buds with two to four scales, or lower leaf buds not much developed.

Leaf scars crowded near twig tips, only a little raised, triangular or somewhat crescent shaped. Bundle scar one. Stipule scars absent.

Fruit an ovoid, glandular-bristly capsule, 4–7 mm long, splitting from the top into four parts; *fruits held erect on long slender stalks, persistent in winter.*

Moist upland woods and balds; common, NC, GA, GSMNP.

| *Leiophyllum buxifolium*
| sand-myrtle

A low, erect shrub. Bark on older stems *shreddy.*

Leaves mostly alternate, simple, *evergreen,* leathery, *small,* 5–15 mm long, elliptical to oblong, entire, bluntly pointed at apex, wedge shaped at base, smooth and *shiny* on upper surface, paler and obscurely dotted beneath; petioles short or almost lacking.

Twigs very slender, nearly circular in cross section.

Buds minute, pointed, with two indistinct scales.

Fruit a dry, smooth capsule, 3–4 mm long, borne in terminal clusters.

Rocky woods and bluffs; Jackson, Macon, and Mitchell Cos., NC; GA; GSMNP.

In *L. buxifolium* var. *prostratum* (Allegany sand-myrtle), stems are lying flat on ground or widely spreading, and leaves are mostly *opposite* or subopposite; capsules *roughish.* High mountains, often in exposed places, NC.

Kalmia (kalmia)

Erect *evergreen* shrubs, rarely small trees.

Leaves alternate or whorled, or sometimes opposite, usually crowded near end of each year's growth, leathery, simple, entire.

Twigs moderate or slender, circular in cross section.

Flower buds much different from leaf buds: borne in clusters at ends of twigs or in axils of upper leaves, covered by many green, overlapping scales coated with gland-tipped hairs. Leaf buds solitary, small, with two green outer scales.

Leaf scars half circular to shield shaped. Bundle scar one, appearing as a transverse line. Stipule scars lacking.

Fruit a globose, five-celled capsule containing many small rounded seeds; in terminal or axillary clusters, *persistent in winter.*

1 Leaves short pointed to long pointed at apex, 5–10 cm long, usually alternate but sometimes opposite or whorled; fruits in terminal clusters
.... (1) *K. latifolia*

1' Leaves blunt or rounded at apex, 2–6 cm long, usually whorled; fruits in axillary clusters .... (2) *K. angustifolia*

FIGURE 129   Mountain-laurel, *Kalmia latifolia*. Twig with clustered leaves and flower bud; terminal cluster of capsules.

FIGURE 129 | 1. *Kalmia latifolia*
mountain-laurel

A crooked shrub or small tree, the branches rigid and spreading, often forming dense thickets. Bark thin, dark reddish brown, divided into long narrow ridges and shredding.

Leaves *evergreen, mostly alternate but crowded near tip of twigs, flat,* leathery, elliptical to broadly lanceolate, 5–10 cm long, entire, short pointed at apex, wedge shaped at base, dark *glossy* green above, yellowish green beneath, glabrous on both sides.

Twigs reddish green and sticky-glandular at first, soon becoming reddish brown and glabrous.

Flower buds *terminal, in distinctive clusters,* 2–3 cm long, covered by glandular scales. Leaf buds in axils of leaves just below flower buds, very small, ovate, flattened, with usually two outer scales.

Flowers saucer shaped, white, rose, or pink, usually with a purple spot, in terminal and axillary clusters; April–June.

Fruit a brown, glandular-hairy, nearly globose woody capsule, 5–7 mm across, in erect *terminal* clusters, persisting in winter.

Wooded hillsides and mountain ravines; common, throughout.

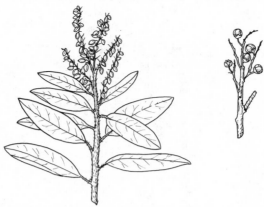

FIGURE 130   Fetter-bush, *Pieris floribunda*. Twig bearing evergreen leaves and clusters of flower buds; portion of stem with persistent capsules.

2. *Kalmia angustifolia*
lambkill, sheep-laurel

Leaves *in whorls of three or opposite,* evergreen, slightly leathery, oblong to elliptical-lanceolate, 2–6 cm long, entire, blunt at apex, rounded to broadly wedge shaped at base, rather dull green above, paler beneath, glabrous on both surfaces, at least at maturity.

Twigs light brown to grayish brown, grayish hairy and often glandular at first, later glabrous.

Flower buds *axillary* (in upper angle between leaves and twig).

Fruit a nearly smooth, flattened globe shaped capsule, 3–4 mm thick, borne in *lateral* clusters on twigs of previous year, persistent all winter.

In dry or wet woods with sterile, acid soils; occasional, NC, GA.

FIGURE 130 | *Pieris floribunda*
fetter-bush, mountain pieris

A dense, *evergreen* shrub.

Leaves alternate, simple, *leathery,* oblong to ovate-lanceolate, 3–8 cm long, entire or *minutely toothed and bristly along margin,* short pointed or long pointed at apex, rounded at base, glabrous or nearly so above, *dotted beneath with black or reddish glands.*

Twigs slender or moderately slender, bristly with sharp, stiff, forward-pointing hairs.

FIGURE 131   Male-berry, *Lyonia ligustrina*. Twig without leaves, twig bearing leaves, sections of twig showing two views of lateral buds.

Flower buds in erect, terminal and axillary clusters, formed in summer and autumn, remaining *naked in winter,* expanding in spring.

Leaf buds very small, pointed, divergent, reddish, with several scales visible. Terminal leaf buds absent.

Leaf scars small, little raised, half circular. Bundle scar one. Stipule scars absent.

Fruit a dry, globe-shaped, five-celled capsule, 4–5 mm long, with a few seeds in each cell; maturing in August to October, usually persisting for a long time.

Exposed slopes and balds at high elevations; occasional, central and southwest mountains of NC; GSMNP.

FIGURE 131 | *Lyonia ligustrina*
| male-berry

An upright, freely branched shrub.

Leaves alternate, deciduous, simple, 3–7 cm long, mostly ovate or obovate, entire or minutely toothed, short pointed or abruptly long pointed at apex, wedge shaped at base, dull green above, paler beneath, usually hairy on both sides, and ciliate along margin.

Twigs slender, *typically yellow* or light brown; glabrous, hairy in lines, or uniformly hairy; sometimes mottled or turning blackish; older twigs often with stringy bark.

Buds slender, flattened, pointed, *pinkish red,* about 3 mm long, usually appressed, sometimes curved, with *two valvate scales* visible. Terminal bud absent.

Leaf scars *shield shaped* to half circular (similar to those of *Oxydendrum*). Bundle scar one. Stipule scars absent.

Fruit a *small globose capsule,* 2–4 mm across, containing numerous seeds; borne in terminal clusters.

This species may be confused with *Vaccinium erythrocarpum* (bearberry), but in that species the leaves are ciliate with glandular hairs, twigs are dark red to green, leaf scars are narrow, resembling those of *Malus* but with only one bundle scar; and fruit is a dark red berry.

Moist or wet woods and streambanks; common, throughout.

Leucothoe (sweetbells)

Evergreen or deciduous shrubs with alternate, simple, toothed leaves.

Twigs moderate to slender, circular in cross section when mature.

Flower buds in *catkinlike or slender, one-sided clusters (racemes), conspicuous in winter.* Leaf buds small, solitary, roundish, with several scales visible. Terminal leaf bud lacking.

Leaf scars raised, crescent shaped to half circular. Bundle scar *one.* Stipules and stipule scars absent.

Flowers and fruits in terminal or axillary racemes. Fruit a dry, flattened globe shaped, five-celled capsule, the style persistent; containing minute irregular seeds; capsules persisting into winter.

1 Leaves thick and leathery, evergreen; flowers and fruits in axillary racemes
 . . . . (1) *L. editorum*
1′ Leaves thin, deciduous; flowers and fruits mostly in terminal racemes:
  2 Leaves tapering to a long, slender point; petioles often more than 3 mm long; flower and fruit racemes usually spreading and curved; capsules 4–6 mm across . . . . (2) *L. recurva*
  2′ Leaves merely short pointed at apex, not tapering to long point; petioles 1–3 mm long; flower and fruit racemes usually upright and straight or nearly so; capsules 3–4 mm across . . . . (3) *L. racemosa*

FIGURE 132	1. *Leucothoe editorum* dog-hobble, upland sweetbells

An *evergreen* shrub with arching and sprawling branches, commonly in thickets.

Leaves *thick, leathery,* elliptical-lanceolate or narrowly ovate, 5–13 cm long, margined with fine, often bristlelike teeth, *tapering to a long point,* wedge shaped to rounded at base, dark green and lustrous above, paler and minutely hairy beneath.

Stems clustered, moderately slender, conspicuously zigzag, reddish to greenish, finely hairy when young.

Flower buds in *short, dense racemes, catkinlike, borne in axils of leaves, exposed in*

FIGURE 132   *Leucothoe.* Left: sweetbells. *L. racemosa.* Leaf, raceme of capsules, two
views of leaf scars and lateral leaf buds. Right: dog-hobble, *L. editorum.*
Sections of twig showing evergreen leaves, catkinlike flower buds, leaf buds,
and raceme of capsules.

winter. Leaf buds set at wide angle to twig, very small, globose, with a few visible
scales.
Capsules flattened globe shaped, 3–3.5 mm long, 5–6 mm thick.
In ravines and along streams; common in moist, acid soils, throughout.

2. *Leucothoe recurva*
mountain sweetbells, fetter-bush

A low shrub with spreading branches.
Leaves deciduous, thin, lanceolate to ovate, 5–10 mm long, finely toothed, *long
pointed* at apex, usually wedge shaped at base, green and glabrous above, hairy
beneath, at least along veins.
Twigs slender, reddish, usually glabrous.
Flower buds *terminal, in long slender racemes,* somewhat resembling catkins, ex-
posed in winter. Leaf buds short, broadly conical, set at wide angle to twig.
Flower and fruit racemes long, one-sided, *curved or nodding.* Capsules flattened-
globose, five-lobed, 4–6 mm across, terminated by persistent style.
Dry rocky woods and thickets, usually at higher elevations; occasional, throughout.

	3. *Leucothoe racemosa*
FIGURE 132	sweetbells, fetter-bush

A shrub with mostly erect branches.

Leaves deciduous, thin, narrowly ovate, 3–9 cm long, finely toothed, *short pointed* or abruptly long pointed at apex, wedge shaped at base, green and glabrous on upper surface, slightly paler and hairy on veins beneath.

Twigs slender, zigzag, reddish or greenish, often reddish brown on side facing light, green beneath, glabrous or nearly so.

Flower buds *terminal,* in slender, *reddish, catkinlike racemes,* present and clearly visible in winter. Leaf buds very small, less than 2 mm long, divergent, roundish, reddish brown, with three or four scales visible.

Flower and fruit racemes one-sided, straight or nearly so, erect or ascending. Capsules roundish, not lobed, 3–4 mm across, tipped with a persistent style.

Swampy thickets and streambanks; Buncombe, Henderson, and Polk Cos., NC; GA.

	Oxydendrum arboreum
FIGURE 133	sourwood

A medium-sized to large tree, but often flowering when much smaller. Mature bark thick, deeply furrowed and blocky.

Leaves alternate, deciduous, simple, lanceolate to elliptical-lanceolate, 8–20 cm long, finely toothed, at least toward apex, or occasionally entire, long pointed at apex, wedge shaped or rounded at base, bright green and lustrous above, paler beneath, glabrous on both surfaces; *sour or bitter to taste;* turning scarlet in late summer or autumn.

Twigs slender to moderate, red, orange brown, or green, often reddish on upper side, green beneath, glabrous, more or less angled below nodes but not grooved as in some species of *Vaccinium.*

Terminal bud absent. Lateral buds *small,* 1–2 mm long, nearly globe shaped but pointed, *appearing partly sunken in bark,* reddish, glabrous, with four to six scales showing.

Leaf scars slightly raised, shield shaped to half circular. Bundle scar one, *crescent shaped to V-shaped.* Stipule scars absent.

Fruit a *small,* five-valved *capsule,* held erect in long, *one-sided,* drooping *clusters,* distinctive, often *persistent in winter.*

Moist woods in valleys and uplands; common, throughout.

	Epigaea repens
FIGURE 134	trailing-arbutus

A small, prostrate shrub with creeping stems rooting at intervals and sending up erect stems.

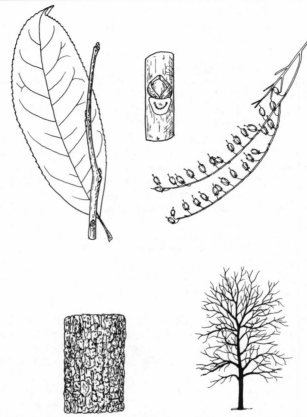

FIGURE 133   Sourwood, *Oxydendrum arboreum*. Leaf and twig; enlarged section of twig
               showing lateral leaf bud, leaf scar, and bundle scar; one-sided clusters of
               fruits, bark of mature tree, silhouette of mature tree.

Leaves *evergreen,* alternate, simple, oblong, 3–8 cm long, entire but hairy mar-
    gined, often wrinkled, blunt or short pointed at apex, rounded or heart shaped
    at base, green and more or less *bristly-hairy* on both surfaces.
Stems slender, almost herbaceous, *covered with coarse reddish hairs.*
Buds conical, solitary or the flower buds multiple, with several scales, the outer pair
    hairy, the inner glabrous and sharply pointed.
Leaf scars linear or absent, the leaves being shed with the peeling bark. Stipules and
    stipule scars absent.

FIGURE 134   Trailing-arbutus, *Epigaea repens*. Branch with evergreen leaves and flower bud.

Fruit a flattened globose capsule, 6–8 mm across, splitting into five parts and exposing fleshy interior; seeds numerous, lustrous, reddish brown.

Sandy, open woods; fairly common, throughout.

FIGURE 135	*Gaultheria procumbens* aromatic wintergreen, checkerberry

A low, *aromatic* shrub with underground stems giving rise to erect, leafy stems less than 20 cm high. Foliage with *wintergreen odor* when crushed.

Leaves alternate, mostly crowded near tip of upright stems, *evergreen,* leathery when mature, elliptical to obovate, 2–5 cm long, obscurely toothed, rounded or blunt at apex, wedge shaped at base, glabrous and glossy on both surfaces.

Erect stems scarcely woody, simple or sparingly branched, very slender, brownish or greenish, glabrous or nearly so.

Buds solitary, flattened globose, with several scales.

Leaf scars very small, only slightly raised, linear or crescent shaped. Bundle scar one.

Fruit a fleshy, globular "berry" (capsule enclosed in fleshy calyx), 8–12 mm across, *red,* persisting through winter; flesh edible but mealy, with wintergreen flavor.

Dry or moist woods; fairly common, NC, GA, GSMNP.

Gaylussacia (huckleberry)

Low shrubs with *resin dotted leaves,* often growing in colonies.

Leaves alternate, ours deciduous or mostly so, simple, entire, covered on lower side with yellow resin droplets or patches (conspicuous in sunlight).

Twigs slender, circular in cross section, commonly pink or red, or greenish on lower side, *not speckled* with warty dots as in some species of *Vaccinium*.

Flower buds and leaf buds separate, larger flower buds with four to seven scales

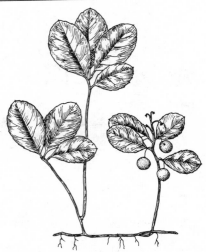

FIGURE 135   Aromatic wintergreen, *Gaultheria procumbens*. Trailing stem and roots; erect stems with leaves and fruits.

visible, often with *yellow resin droplets near tip,* smaller leaf buds with mostly two to four scales showing. Terminal bud absent.

Leaf scars small, crescent shaped to half circular. Bundle scar one. Stipule scars absent.

Fruit berrylike, mostly black, containing *ten seedlike nutlets.*

1   Leaves glossy green on upper side; young twigs with gland-tipped hairs; leaf buds with four or five scales exposed . . . .(1) *G. dumosa*

1′ Leaves dull green or yellowish green on upper side; young twigs with gland-less hairs or without hairs; leaf buds with two (occasionally three) scales exposed;

   2   Young growth sticky, the leaves bearing glandular dots or patches on both surfaces; leaf buds usually less than 2 mm long; flower buds with five to seven scales visible . . . .(2) *G. baccata*

   2′ Young growth not sticky, the leaves bearing glandular dots only on lower surface; leaf buds 2–4 mm long; flower buds with four scales visible, the shortest scale at least three-fourths as long as bud . . . . (3) *G. ursina*

FIGURE 136 | 1. *Gaylussacia dumosa*
                   | dwarf huckleberry

Leaves deciduous or partly evergreen, somewhat leathery, oblanceolate to obovate, 2–4 cm long, entire, rounded at apex (midrib projecting as a short, stiff point), wedge shaped at base, *glossy green above*, slightly paler and *glandular-hairy* with scattered resin droplets *beneath*.

Twigs glandular-hairy, light reddish brown to gray.

Flower buds larger than leaf buds, light brown to reddish, several scaled, sprinkled with resin droplets. Leaf buds ovoid, somewhat divergent, with *four or five scales exposed*.

Flower and fruit stems with *persistent leaflike bracts*, elliptical to oblong, 5–12 mm long, resin dotted, mostly on lower side.

Fruit black, shiny, 7–10 mm across, edible but rather tasteless.

Sandy or dry woods at lower elevations, southwest mountains of NC and adjacent SC.

	2. *Gaylussacia baccata*
FIGURE 136	black huckleberry

Leaves elliptical to oblong or oblong-lanceolate, 2–6 cm long, entire, short pointed to rounded at apex, wedge shaped at base, yellowish green above, paler beneath, dotted on *both sides* with shiny, yellowish resin droplets, the leaves sticky when young (glandular dots or patches on upper surface may be widely scattered or nearly lacking late in season).

Twigs pink or red to green, often pink above and green beneath, hairy with fine, whitish curly hairs.

Buds of two kinds: flower buds about 3 mm long, pink, brown, or green, with five to seven visible scales and commonly with yellow resin dots near tip; leaf buds 1–2 mm long, same color as twigs, with *two or three visible scales,* the latter pointed but not with prominent spinelike points as in some species of *Vaccinium*.

Fruit black, lustrous, 6–8 mm across, sweet and edible.

Dry, open woods; common at lower elevations, throughout.

	3. *Gaylussacia ursina*
FIGURE 136	buckberry, bear huckleberry

Leaves elliptical to elliptical-oblanceolate, 5–13 cm long, entire, short pointed to long pointed at apex, tip of midrib protruding (mucronate), wedge shaped to rounded at base, greenish on both surfaces, glabrous above, hairy beneath, at least along midrib and principal lateral veins; *resin dotted only on lower side;* petioles golden brown–hairy.

Young twigs pink or tan to green, covered with golden brown and sometimes whitish hairs; older twigs light brown to reddish brown.

Buds mostly 2–4 mm long, ovoid to oblong, bluntly pointed, pink to green, glabrous or thinly hairy. Flower buds with about *four scales exposed,* often with resin droplets on scales, commonly narrowed at base but not stalked. Leaf buds with two scales visible.

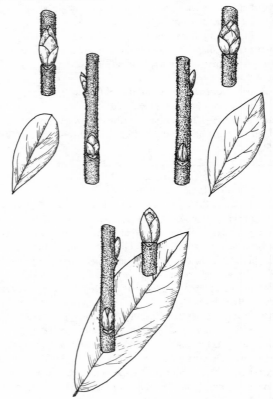

FIGURE 136   *Gaylussacia*. Upper left: dwarf huckleberry, *G. dumosa*. Leaf, enlarged
flower bud, enlarged leaf buds. Upper right: black huckleberry, *G. baccata*.
Enlarged flower bud, enlarged leaf buds, and leaf. Below: buckberry, *G.*
ursina. Enlarged flower bud, enlarged leaf buds, and leaf.

Fruit black, shiny, 7–10 mm across, sweet.
Moist woods at middle and upper elevations, southwest NC and adjacent SC; GA;
GSMNP.

Vaccinium (blueberry)

Deciduous or evergreen shrubs, erect or occasionally trailing.

Leaves alternate, simple, entire or very finely toothed, teeth sometimes ending in
minute hairs; *leaves not dotted with resin droplets* as in *Gaylussacia*.

Twigs slender, often green or red, and often speckled with many raised, whitish dots.

Buds small, with two to several or many visible scales; flower buds and leaf buds commonly separate. Terminal bud absent, twig tips dying back to minute stub at side of topmost lateral bud.

Leaf scars *small, raised,* half circular to triangular or crescent shaped. Bundle scar only *one.* Stipule scars absent.

Fruit a *many-seeded berry,* edible or not.

1 Plants creeping, rooting at nodes, the fruiting stems erect or ascending; leaves evergreen, 5–18 mm long, entire.................(1) *V. macrocarpon*
1′ Plants upright, not creeping; leaves longer than 18 mm:
  2 Leaves thick and leathery, evergreen or tardily deciduous, narrowly elliptical to nearly circular, glossy, entire or minutely toothed; buds small, globose, 1–2 mm long, with several reddish brown to reddish purple scales exposed ....(2) *V. arboreum*
  2′ Leaves not thick and leathery, deciduous:
    3 Buds about as thick as long, 1–2 mm long, roundish but with pointed tip, usually set at about forty-five-degree angle to twig; visible bud scales two to four; leaves whitish or pale bluish beneath ....(3) *V. stamineum*
    3′ Buds longer than thick, set at narrower angle to twig:
      4 Buds distinctly flattened and appressed, mostly 3–6 mm long, gradually tapering to tip but not long pointed; bud scales two, valvate; leaves very finely toothed, the teeth ending in hairlike points, pale and shiny on lower surface.....................(4) *V. erythrocarpum*
      4′ Buds not distinctly flattened and appressed, some or all with the outer pair of scales extended into spinelike points; visible bud scales usually more than two, not valvate; leaves toothed or entire:
        5 Young twigs evenly and densely hairy, obscurely warty dotted; leaves hairy beneath:
          6 Leaves mostly more than 4 cm long, entire or minutely toothed, usually thinly hairy to glabrous on upper side; tall shrub; lower elevations, occasional.......................(5) *V. atrococcum*
          6′ Leaves mostly less than 4 cm long, entire, densely hairy on upper side; low shrub; high mountains, rare .... (6) *V. hirsutum*
        5′ Young twigs glabrous or hairy only in longitudinal grooves or bands, conspicuously speckled with whitish warty dots; leaves glabrous beneath or hairy merely on veins:
          7 Low shrub, 20–60 cm in height but usually less than 50 cm, freely branched near the ground; twigs green or yellowish green; leaves entire or faintly toothed ....(7) *V. vacillans*
          7′ Tall shrubs, 1–5 m in height, little branched until more than 50 cm tall; twigs green or often bright red:

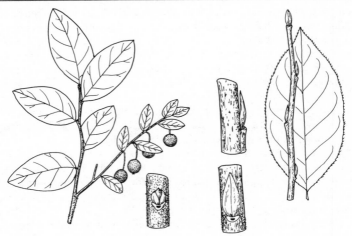

FIGURE 137   *Vaccinium.* Left: deerberry, *V. stamineum.* Twig with leaves, fruits, and leafy bracts; enlarged bud and leaf scar. Right: bearberry, *V. erythrocarpum.* Twig, leaf, and two enlarged views of lateral buds.

8   Leaves usually minutely toothed .... (8) *V. constablaei*
8′ Leaves entire............................(9) *V. corymbosum*

1. *Vaccinium macrocarpon*
cranberry

A low, *trailing* shrub.

Leaves *evergreen,* leathery, elliptical, 5–18 mm long, entire, flat or with margin slightly rolled under, blunt or rounded at apex, wedge shaped to rounded at base, glossy dark green above, paler beneath.

Stems creeping, slender, light brown to reddish brown, glabrous or somewhat hairy, rooting at nodes and sending up fruiting branches 10–20 cm high.

Fruit 10–15 mm across, red or dark red, very acid, sometimes persisting into winter.

Boggy places; occasional, NC.

2. *Vaccinium arboreum*
sparkleberry, farkleberry

A large shrub or small tree.

Leaves evergreen or tardily deciduous, more or less *leathery,* elliptical to obovate, 2–7 cm long, entire or minutely toothed, somewhat rounded to short pointed at apex, broadly wedge shaped at base, glabrous and *glossy* on upper side, glabrous or hairy beneath.

Twigs *reddish brown* to grayish brown, glabrous or thinly hairy with curly white hairs.

Buds all of one kind, small, 1–2 mm long, nearly *globose,* divergent, reddish brown to reddish purple, glabrous, with four to six scales exposed.

Fruit *black,* 5–8 mm across, dry, inedible.

Dry, sandy or rocky woods at lower elevations; southwest NC, SC, GA, GSMNP.

FIGURE 137	3. *Vaccinium stamineum* deerberry, squaw-huckleberry

A many-branched, erect shrub.

Leaves having *two forms,* those on vegetative twigs oblong or elliptical to oblanceolate, 3–10 cm long, entire, short pointed or long pointed at apex, wedge shaped to slightly heart shaped at base, dull green and usually glabrous above, *bluish whitened beneath;* leaves (leaflike bracts) on flower and fruit stems typically much smaller, some only 5–20 mm long, and somewhat different in shape.

Twigs yellowish brown or reddish brown to greenish, *not warty dotted,* finely curly hairy or glabrous.

Buds all about alike, *widely divergent,* small, 1–2 mm long, broadly ovoid to globose, pointed at tip, with usually three (two to four) reddish scales visible, the scales often slender pointed but not prolonged into spinelike points as in blueberries.

Berries 7–15 mm across, yellowish, pink or greenish to blue, glaucous, sometimes edible but more often not.

Open, rather dry woods; common, throughout.

FIGURE 137	4. *Vaccinium erythrocarpum* bearberry, mountain-cranberry

An upright shrub, often forming colonies.

Leaves deciduous, thin, lanceolate to ovate-lanceolate, 3–8 cm long, *minutely toothed and ciliate,* the cilia tipped with glands, long pointed or short pointed at apex, broadly wedge shaped to heart shaped at base, dull green above, paler and *lustrous beneath,* hairy along veins on both surfaces.

Twigs distinctly grooved downward from the nodes and hairy in the grooves, faintly warty dotted, dark red or brownish red to green, often reddish on side facing light, green on lower side; older bark either close or peeling.

Buds *all about same size and shape,* mostly 3–6 mm long, slender, pointed, *flattened, closely appressed* to twig, brownish red to green; exposed bud scales only *two, valvate.*

Fruit solitary, in axils of leaves, *dark red,* 4–6 mm across, acid and rather tasteless.

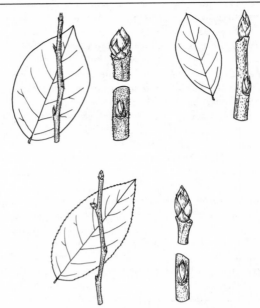

FIGURE 138   *Vaccinium.* Upper left: black highbush blueberry, *V. atrococcum.* Leaf; twig;
enlarged flower bud, leaf scar, and stub of end of season's growth; enlarged
leaf bud, both buds showing spinelike extensions of lower scales. Upper
right: lowbush blueberry, *V. vacillans.* Leaf; enlarged section of twig
showing flower bud and leaf bud. Below: highbush blueberry, *V. constablaei.*
Leaf; twig; enlarged end of twig showing last flower bud, leaf scar, and stub
of season's growth; enlarged leaf bud showing lower scales extended into
spinelike points.

May be confused with male-berry (*Lyonia ligustrina*) but that species has pinkish
    buds, twigs mostly yellow, and fruit a small globose capsule.
Moist woods, mostly at higher elevations; infrequent, throughout.

   Cross-breeding is common among the true blueberries, and this often makes identification of species uncertain. Some of our plants are almost impossible for anyone
not a specialist in the group to name. Consequently, there may be individual plants
that do not fit any of the following descriptions closely.

| | 5. *Vaccinium atrococcum* |
| FIGURE 138 | black highbush blueberry |

A tall shrub.
Leaves elliptical to elliptical-lanceolate, 4–8 cm long, minutely toothed or entire,

short pointed at apex, usually wedge shaped at base, dark green and sparsely hairy or glabrous above, pale green and *woolly beneath.*

Twigs *densely hairy* with pale curly hairs, warty dotted and reddish or greenish under the curly hairs.

Buds red, more or less *hairy,* some or all scales tapering to a long sharp point. Leaf buds slender, 2–3 mm long, with about four scales visible; flower buds more plump than leaf buds, with six to eight scales exposed.

Fruit 5–10 mm across, black, without bloom, juicy.

Both dry and wet woods; occasional, throughout.

| 6. *Vaccinium hirsutum*
| hairy blueberry

A low shrub.

Leaves ovate to elliptical-oblong, mostly 2–5 cm long, entire, short pointed at apex, broadly wedge shaped at base, dull green above, paler beneath, *hairy on both sides.*

Twigs evenly and rather densely *hairy,* indistinctly warty dotted beneath the hairs.

Flower and leaf buds separate, the flower buds larger and with more scales than the leaf buds; bud scales hairy, some or all prolonged into slender points.

Fruit 6–7 mm across, purplish black, glandular-hairy, edible.

Woods at higher elevations; Cherokee and Graham Cos., NC; GA; GSMNP.

FIGURE 138 | 7. *Vaccinium vacillans*
| lowbush blueberry

A *low,* finely branched shrub.

Leaves broadly elliptical to elliptical-ovate or elliptical-obovate, mostly 2–4 cm long, one-half to two-thirds as wide, entire or nearly so, blunt or short pointed at apex, the *midrib protruding* beyond apex as a stiff, slender point (mucro); dull green and glabrous above; lower surface usually glaucous and glabrous, or sometimes hairy only on veins.

Twigs slender or very slender, grooved above buds and below tips of leaf scars, *distinctly speckled with warty dots, green or yellowish green,* or sometimes reddish on upper side, hairy only in narrow bands, otherwise glabrous.

Buds red or reddish brown to golden brown, *the scales,* at least some, *extended into spinelike points.* Leaf buds 1–2 mm long, with two to four scales showing; flower buds more plump than leaf buds, about 3 mm long, with five to eight scales visible.

Fruit 5–10 mm across, blue, glaucous, sweet and edible.

Dry or moist woods; common, throughout.

| | 8. *Vaccinium constablaei* |
| FIGURE 138 | highbush blueberry |

A tall shrub.

Leaves elliptical, elliptical-lanceolate, or lanceolate, mostly 4–8 cm long, usually less than 3 cm wide, *very finely toothed* to entire, long pointed at apex, mostly wedge shaped at base, *glaucous,* at least when young, glabrous on both sides or hairy on midrib and principal veins beneath.

Twigs green or yellowish green to deep red, often reddish on upper side, green on lower, *roughened with whitish warty dots,* hairy in long grooves above and below nodes, but otherwise glabrous.

Buds red or reddish brown to golden brown, glabrous, the scales, at least the lower pair, *prolonged into slender points.* Leaf buds slender, with three to five scales ghowing; flower buds more plump than leaf buds, with six to nine visible scales, the latter often with ragged edges.

Fruit 5–10 mm across, bluish black, glaucous, sweet and juicy.

Wooded slopes at higher elevations; common, throughout.

| | 9. *Vaccinium corymbosum* |
| | northern highbush blueberry |

Closely resembles *V. constablaei,* but its leaves are usually *entire* and short pointed at apex.

Streambanks and moist woods; Haywood, Mitchell, and Yancey Cos., NC; GA.

Ebony Family (Ebenaceae)

Trees or shrubs with alternate, deciduous, simple, entire leaves; flowers small, staminate and pistillate flowers generally on different plants, solitary or few along twigs; fruit an edible berry. Represented in our area by one species.

| | *Diospyros virginiana* |
| FIGURE 139 | persimmon |

A small to medium-sized tree. Bark of trunk thick, nearly black, *deeply divided into small, squarish blocks.*

Leaves alternate, deciduous, simple, elliptical or oblong, mostly 6–16 cm long and 3–7 cm wide, entire, long pointed or short pointed at apex, mostly *rounded at base,* dark green and glossy above, paler beneath, very smooth in texture, midrib and veins often strikingly reddish or brownish on lower side.

FIGURE 139   Persimmon, *Diospyros virginiana*. Leaf, fruit, twig, enlarged bud and leaf scar, mature bark.

Twigs slender to moderate, slightly zigzag, *purplish* to dark brown or grayish brown; pith often spongy or chambered in older twigs.

Terminal bud absent. Lateral buds solitary, ovoid, 2–4 mm long, pointed, *dark reddish to purplish;* visible bud scales two, the first scale as long as the bud and much overlapping the second.

Leaf scars raised, semicircular to semielliptical. Bundle scar *one,* appearing as *a transverse curved line.* Stipules and their scars absent.

Fruit a fleshy, plumlike berry, yellowish brown to purplish, containing large flat seeds, maturing in autumn and sometimes lasting into winter; flesh puckery at first ("furs the mouth") but sweet when thoroughly ripe after frosts.

Apt to be confused with black tupelo (*Nyssa sylvatica*), but that species has leaves wedge shaped at base, three distinct bundle scars, and a terminal bud with several scales showing.

Dry, deciduous woods, roadsides, old fields, clearings; infrequent at lower elevations, throughout.

Sweetleaf Family (Symplocaceae)

A family consisting of a single genus: trees or shrubs with alternate, simple, deciduous, entire or finely toothed leaves; flowers small, in short axillary clusters; fruit dry, drupelike. Only one species occurs in our area.

	Symplocos tinctoria
FIGURE 140	sweetleaf

A shrub or small tree.

Leaves alternate, leathery, deciduous or somewhat persistent, simple, oblong to narrowly elliptical, 8–15 cm long, the margin entire, wavy, or very finely toothed; long pointed at apex, wedge shaped at base, yellowish green above, slightly paler beneath, hairy to nearly glabrous on both surfaces; *sweet to taste;* some leaves crowded toward tip of well-developed twigs.

Twigs moderate to slender, brown to grayish brown, covered with brown to grayish hairs, prominently angled below nodes; branching often in false whorls; *pith chambered.*

Terminal bud present, conical 2–4 mm long, with about four keeled scales visible, the scales more or less hairy; several buds (and leaf scars) clustered at twig tip, larger terminal bud and three or more smaller leaf buds. Lateral buds of *two distinct kinds: (a) very small,* hairy, few-scaled *leaf buds,* commonly appearing partly sunken in bark, and (*b*) *globe-shaped flower buds* with five or more scales showing; buds sometimes superposed.

Leaf scars only a little raised, broadly crescent shaped to half circular. Bundle scar *one,* crescent shaped, open side up, or occasionally divided. Stipule scars absent.

Fruit a dry, green, ovoid drupe, 8–12 mm long, in short axillary bunches along twigs, maturing in late summer or early autumn.

Rich woods, sandy soils along streams; infrequent, NC, SC, GA.

Storax Family (Styracaceae)

Shrubs or small to large trees of the South; leaves alternate, simple, deciduous, entire or finely toothed; flowers small to medium sized, bell shaped, white or pinkish, often showy; fruit dry, nutlike, elongated, two- to four-winged. One species described for our area.

FIGURE 140   Sweetleaf, *Symplocos tinctoria*. Leaf, detailed sections of twig showing terminal bud, lateral leaf bud, lateral flower bud (superposed), and chambered pith.

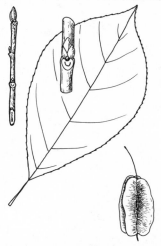

FIGURE 141   Carolina silverbell, *Halesia carolina*. Twig with terminal bud; leaf; detailed section of twig showing leaf scar, bundle scar, and superposed buds; winged fruit.

	Halesia carolina
FIGURE 141	Carolina silverbell

A shrub or small to large tree. Young trunks and branches with *bark longitudinally striped*.

Leaves alternate, deciduous, simple, broadly elliptical to obovate, 10–17 cm long, finely toothed, long pointed at apex, wedge shaped at base, dark yellowish green and glabrous above, slightly paler and thinly hairy or glabrous beneath.

Twigs slender, light brown to light reddish brown, hairy at first but soon glabrous; bark *silky-shreddy* on twigs two years old and older. Pith *finely chambered,* the chambers hollow.

Terminal bud present, at least on some twigs. Buds ovoid to conical, pointed, 2–5 mm long, *reddish,* with two to four scales visible. Lateral buds often superposed, a very small bud at base of the main bud.

Leaf scars large in relation to size of twig, raised, half circular to nearly circular, often notched or split at top. Bundle scar *one, U-shaped or V-shaped,* or sometimes several scars crowded closely together. Stipule scars absent.

Fruit a dry oblong drupe, 3–4 cm long, with *four prominent wings.*

Rich, moist woods; common, throughout.

Olive Family (Oleaceae)

A large family of trees or shrubs with opposite, deciduous or half-evergreen, simple or pinnately compound leaves; flowers generally small, sometimes showy, often in clusters, usually bisexual, or staminate and pistillate flowers on separate plants (*Fraxinus*); fruit a samara with a single wing, two-celled capsule, or fleshy one-seeded drupe; ash, fringetree, lilac, and others.

Fraxinus (ash)

Trees with *opposite,* deciduous, *pinnately compound* leaves; leaflets five to eleven, toothed or not.

Twigs moderately slender to stout, stiff, more or less *flattened at nodes.*

Terminal bud present, fairly large, with two or three pairs of visible scales appearing *scurfy or granular.* Lateral buds similar but smaller, the first pair set directly at base of terminal bud.

Leaf scars opposite, *large, nearly circular to crescent shaped.* Bundle scars *numerous,* close together, often more or less fused, in a U-shaped or C-shaped line. Stipules and stipule scars lacking.

Fruit a dry, seedlike samara, *prominently winged at tip,* in loose drooping clusters in axils of leaves. Pollen-bearing and seed-producing flowers borne on separate trees, and hence not all trees bearing fruit.

Summer Key

1 Young twigs and petioles glabrous or nearly so . . . . (1) *F. americana*
1' Young twigs and petioles hairy or woolly:
  2 Young twigs and petioles covered with soft, short hairs:
    3 Leaflets very pale or whitish beneath. . . . .(2) *F. americana* var. *biltmoreana*
    3' Leaflets light green or rusty green beneath . . . .(3) *F. pennsylvanica*
  2' Young twigs and petioles densely hairy with matted wool
    . . . . (4) *F. profunda*

Winter Key

1 Leaf scars with the upper margin straight or convex (but often slightly indented by bud) . . . .(3) *F. pennsylvanica*
1' Leaf scars with the upper margin deeply concave, making leaf scar U-shaped or V-shaped:
  2 Twigs glabrous or mostly so . . . . (1) *F. americana*
  2' Twigs hairy or woolly:
    3 Twigs densely covered with woolly hairs . . . . (4) *F. profunda*
    3' Twigs merely short-hairy . . . .(2) *F. americana* var. *biltmoreana*

FIGURE 142   *Fraxinus.* Above: white ash, *F. americana.* Enlarged sections of twig
showing terminal bud complex, two lateral buds, and two forms of leaf
scars; winged fruit; and pinnately compound leaf. Below: green ash, *F.
pennsylvanica.* Pinnately compound leaf, winged fruit, and sections of twig
showing terminal bud complex, lateral buds, and two forms of leaf scars.

FIGURE 142	1. *Fraxinus americana* white ash

A large tree. Bark of trunk grayish brown, roughened by network of crisscrossing ridges with diamond-shaped furrows between them.

Leaflets five to nine, usually seven, ovate to oblong-lanceolate, 7–18 cm long, toothed or entire, short pointed or long pointed at apex, broadly wedge shaped or rounded at base, dark green above, *whitish or very pale beneath;* leaflets long stalked, the stalks without wings.

Twigs greenish when young, becoming brownish to gray with age, nearly or entirely glabrous, with many pale lenticels.

Terminal bud broadly ovoid, blunt, dark brown, with two or three pairs of scales visible, these having sharp, abrupt point and granular or scurfy surface. Lateral buds smaller, set in a deep notch in the leaf scar, the first pair of buds located directly at base of terminal bud.

Leaf scars half circular to shield shaped in outline, but *deeply indented at top.*

Fruits produced only on pistillate trees: clusters of lance-shaped keys, each with wing 3–7 mm wide and extending only a little way on seed body.

Rich, moist woods; fairly common, throughout.

	2. *Fraxinus americana* var. *biltmoreana* Biltmore ash

This tree is similar to the typical variety, except that twigs, petioles, and stalks of leaflets are finely hairy. Low woods; occasional, throughout.

FIGURE 142	3. *Fraxinus pennsylvanica* green ash

A tree, smaller than white ash. Bark grayish brown, with a close-fitting network of narrow ridges and diamond-shaped furrows.

Leaflets usually seven (seven to eleven), oblong-lanceolate to elliptical, 8–15 cm long, entire, wavy margined, or toothed, long pointed to blunt at apex, wedge shaped to rounded at base, medium green above, paler and hairy beneath; *petioles covered with short, velvety hairs;* leaflets short stalked, the stalk more or less winged.

Twigs greenish gray to gray, *velvety-hairy* to glabrous.

Terminal bud broadly ovoid, rounded at tip, reddish brown to brown, with two pairs of scales showing; bud scales with granular or scurfy surface. Lateral buds similar but smaller, set above leaf scar or in small notch of leaf scar.

Leaf scars semicircular to nearly circular, *the upper margin straight across, only slightly concave,* or convex but often indented by bud.

Fruits produced only on pistillate trees: clusters of lance-shaped keys, each with a
    wing less than 6 mm broad, pointed or slightly notched at tip, and extending
    about halfway down sides of extremely slender seed body.
Low woods; Madison Co., NC; infrequent in GSMNP.

| | 4. *Fraxinus profunda*
| | pumpkin ash

A large tree.
Leaflets seven to nine, elliptical to ovate-lanceolate, 10–25 cm long, entire but
    sometimes wavy margined, *long pointed* at apex, straight across or rounded and
    often unequal at base, dark yellowish green and lustrous above, *green beneath*
    and softly hairy to nearly glabrous; *petioles usually densely hairy.*
Twigs grayish brown, *heavily coated with woolly hairs.*
Terminal bud rounded, dome shaped, rusty brown to dark brown (occasionally
    nearly black). Lateral buds set in upper margin of leaf scar.
Leaf scars with the upper margin *strongly curved downward* to make a U or V.
Fruits produced only on pistillate trees: clusters of spear-shaped to elliptical keys,
    large, 5–8 cm long, with *wings 6–12 mm broad* and extending to middle of
    seed body or beyond.
Low woods and soils subject to flooding; Ashe, Swain Cos., NC.

| | *Chionanthus virginicus*
FIGURE 143 | fringetree

A large shrub or small tree, closely related to the ashes (*Fraxinus*).
Leaves opposite or subopposite, deciduous, *simple,* elliptical to obovate, 7–20 cm
    long, *entire,* short pointed or blunt at apex, wedge shaped at base, dark green
    and glabrous above, paler and glabrous to densely hairy beneath.
Twigs moderate to stout, slightly enlarged and flattened at nodes, somewhat similar
    to twigs of ash, light brownish gray to olive gray, hairy to nearly glabrous, cov-
    ered with raised warty lenticels.
Terminal bud present, 2–4 mm long, orange brown to grayish brown, glabrous or
    nearly so, enclosed by three to five pairs of *keeled,* pointed scales. Lateral buds
    smaller, rounded, superposed.
Leaf scars large in relation to size of twig, opposite or subopposite, raised, half
    circular to circular but flattened at top. Bundle scar *one, appearing as a U-
    shaped line* of dots not or scarcely distinct individually. Stipule scars absent, and
    opposing leaf scars not connected by lines.
Fruit a one-seeded drupe, resembling a blue olive, ripening in late summer or early
    autumn and usually dropping before winter sets in.
Open woods along streams, in moist ravines, and on ridgetops at lower elevations,
    throughout.

FIGURE 143   Fringetree, *Chionanthus virginicus*. Leaf, twig, and detailed section of twig showing leaf scar, U-shaped bundle scar, and superposed buds.

Ligustrum (privet)

Semievergreen shrubs or rarely small trees.

Leaves *opposite,* simple, *long persistent, often remaining green through winter,* entire.

Twigs slender, straight, rounded or four-angled below nodes, grayish brown to olive green, glabrous to densely short-hairy.

Terminal bud present, with two or three pairs of overlapping scales visible. Lateral buds small, blunt, solitary or superposed.

Leaf scars opposite, small, raised, half circular; *opposing leaf scars not meeting around twig and not connected by a distinct transverse line.* Bundle scar *one.* Stipule scars absent.

Fruit a small, hard, berrylike drupe, typically blue black, in terminal clusters, often persistent.

Introduced from China and Japan, persisting at old homesites and escaping into woods, thickets, and waste places.

A frequent escape in our area is *L. sinense* (Chinese privet) (Figure 144). Twigs and leaf petioles densely hairy; leaves elliptical to ovate, 3–7 cm long; drupes blue, 6–7 mm long, 4 mm across.

| | *Syringa vulgaris* |
| FIGURE 145 | common lilac |

A large shrub.

Leaves *opposite,* deciduous, simple, more or less *heart shaped,* 5–10 cm long, 3–7

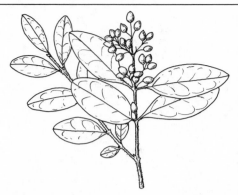

FIGURE 144   Chinese privet, *Ligustrum sinense*. Twigs bearing leaves and fruit clusters.

FIGURE 145   Common lilac, *Syringa vulgaris*. Leaf, twig showing paired lateral buds at end of twig, lower lateral buds and leaf scars.

cm wide, *entire, long pointed* at apex, dark green above, only slightly paler beneath, glabrous throughout.

Twigs slender to moderately stout, stiff, olive green, brown, or gray, glabrous.

Terminal bud usually absent, resulting in forking of twigs. Lateral buds divergent, broadly ovoid, 4–8 mm long, bluntly pointed, greenish to dark red or dark reddish brown, glabrous; bud scales three or four pairs, fleshy.

Leaf scars opposite, crescent shaped to half circular; opposing leaf scars widely separated, not nearly meeting around twig. Bundle scar *one,* made up of many scars more or less united, forming a transverse line. Stipule scars none.

Fruit a small leathery capsule, 10–15 mm long, borne on twigs of previous year.

Native of southeastern Europe, persisting in a few places after cultivation.

Dogbane Family (Apocynaceae)

A large family, mostly herbaceous and tropical, but one species in our area slightly woody; sap milky; leaves chiefly opposite, simple, ours evergreen; flowers rather large, solitary in axil of leaves; fruit a follicle.

FIGURE 146	*Vinca minor* periwinkle

A *trailing*, slightly woody plant.

Leaves opposite, simple, *evergreen*, rather leathery, elliptical-ovate, 20–45 mm long, entire, short pointed at apex, wedge shaped at base, green and *lustrous* above, paler green beneath, glabrous or essentially so on both sides.

Stems glabrous, with *milky sap* when young; horizontal stems creeping, sending up flowering stems 6–15 cm high, forming a dense mat.

Fruits two, erect or spreading, podlike follicles 1–3 cm long; plants seldom fruiting.

Native of southern Europe, occasionally escaped from cultivation into roadsides, waste places, open woods.

Vervain Family (Verbenaceae)

Woody plants or herbs with mostly four-angled stems and opposite leaves, these simple and coarsely toothed in our species; flowers usually small, commonly in small clusters; fruit drupelike or berrylike.

Callicarpa americana beautyberry, French-mulberry

A small to medium-sized shrub.

Leaves opposite, deciduous, simple, elliptical-ovate to ovate-oblong, 7–15 cm long, coarsely toothed, long pointed at apex, wedge shaped at base, more or less hairy above, *whitish-woolly beneath*.

Twigs slender, slightly four-angled at the nodes, grayish brown; *young twigs scurfy*, at least near buds, with minute tufts of branched hairs, visible under magnifying lens.

Buds superposed and commonly stalked, *the larger bud naked*, the smaller with two nearly valvate scales.

Leaf scars opposite, broadly crescent shaped. Bundle scar one, or numerous scars crowded in a U-shaped group and appearing as one. Stipule scars lacking.

FIGURE 146   Periwinkle, *Vinca minor*. Trailing stem and upright stem with evergreen leaves.

Fruit globose, fleshy, berrylike, 3–6 mm across, *rose colored to violet,* containing two to four seeds; densely clustered in axils of leaves.

Moist woods and thickets; Graham and Transylvania Cos., NC; Greenville and Pickens Cos., SC; GSMNP.

Nightshade Family (Solanaceae)

Mostly herbs, only a few members woody; leaves alternate, simple, ours deciduous, entire or lobed; flowers solitary or in various sorts of clusters; fruit a berry or capsule.

FIGURE 147	*Solanum dulcamara* bitter nightshade, bittersweet

A *climbing or trailing* plant, *without tendrils,* woody only near base.

Leaves alternate, deciduous, simple, ovate or heart shaped, 6–11 cm long, *entire or deeply lobed at base,* long pointed at apex, dark green, glabrous or essentially so, with unpleasant odor when crushed.

Stems often dying back in winter, light gray or greenish, two-ridged downward from nodes; pith spongy or *stems hollow.*

Buds rounded, densely white-hairy, with about four indistinct scales.

Leaf scars raised, circular but flattened at top. Bundle scar one, or occasionally broken into three small scars.

FIGURE 147   Bitter nightshade, *Solanum dulcamara*. Fruit cluster, two forms of leaves, and stem.

Fruit a crimson berry, 8–10 mm across, in drooping clusters.
Native of Eurasia, rarely escaped into thickets and clearings.

Figwort Family (Scrophulariaceae)

Mostly a family of herbs; no native trees or shrubs, but one naturalized tree species; leaves opposite, simple, deciduous, entire; flowers showy, deep purple, fragrant; fruit a woody capsule containing many seeds.

FIGURE 148	*Paulownia tomentosa* royal paulownia, princess-tree

A small or medium-sized, widely spreading tree with violet flowers.
Leaves *opposite* (two at a node, not three as usual in catalpa), deciduous, simple, large, ovate to nearly circular, mostly 15–30 cm long but sometimes longer,

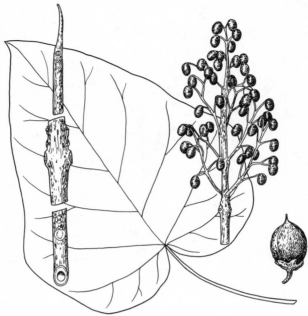

FIGURE 148   Royal paulownia, *Paulownia tomentosa*. Leaf, detailed sections of twig
showing dead tip, superposed buds, opposite leaf scars, and bundle scars in
a nearly closed ring; terminal cluster of flower buds; mature capsule.

entire or slightly lobed, short pointed at apex, heart shaped at base, dark green
and hairy above, lighter green and *soft-velvety beneath.*

Twigs stout, brittle, markedly flattened at nodes, olive brown to dark brown, mostly
glabrous but hairy near tip, around buds, and along upper edge of leaf scar;
lenticels pale, prominent, elongated longitudinally; pith *chambered or hollow.*

Terminal leaf bud absent. Lateral leaf buds *superposed* (one above another), blunt,
partly sunken in bark, projecting at right angle to twig, pale-hairy. Flower buds
about 1 cm long, in erect terminal clusters, exposed in winter.

Leaf scars opposite, vertically elliptical to almost circular; opposing leaf scars not
connected by a transverse line. Bundle scars numerous, arranged in a nearly
closed ring. Stipule scars absent.

Flowers large, showy, violet, fragrant, in upright clusters terminating stout, hairy
twigs; March–May.

Fruit a *woody, beaked, ovoid capsule,* 3–4 cm long, borne in terminal clusters,
mature in autumn, the husks persisting in winter after small winged seeds have
been shed.

Native of China, escaped from cultivation to roadsides and waste places.

Bignonia Family (Bignoniaceae)

Trees or woody vines; leaves mostly opposite or whorled, usually deciduous, simple or pinnately compound; flowers large, showy, in clusters; fruit a dry capsule containing many large seeds with flat wings; trumpet creeper, cross vine, catalpa.

FIGURE 149	*Campsis radicans* trumpet creeper, trumpet vine

A large vine, sometimes creeping or shrubby but usually climbing, mainly by *aerial rootlets.*

Leaves opposite, deciduous, pinnately compound; leaflets seven to fifteen, ovate, 3–7 cm long, long pointed at apex, rounded or wedge shaped at base; margin sharply and *coarsely* toothed, the teeth often prolonged; dull yellowish green and glabrous on upper surface, paler and hairy along veins beneath.

Stems pale yellowish or *straw colored,* glabrous, climbing or trailing, often with aerial rootlets *in two rows or patches below nodes;* tendrils absent.

Buds small, yellowish, glabrous, with two or three pairs of scales.

Leaf scars opposite, low, half circular or shield shaped. Opposite leaf scars connected by a transverse hairy ridge. Bundle scar one, C-shaped, open side up. Stipule scars lacking.

Fruit a podlike capsule, 10–15 cm long, containing winged seeds.

Woods, fencerows, waste places; infrequent, throughout.

	Bignonia capreolata cross vine

A vine climbing by means of *coiling tendrils.*

Leaves opposite, deciduous or half evergreen, compound, each leaf consisting of usually *two leaflets* and, between them, branched tendril ending in adhesive disks. Leaflets oblong or elliptical, 6–15 cm long, entire, short pointed or long pointed at apex, somewhat heart shaped at base, dark green and glabrous above or hairy along midrib, glabrous beneath.

Stems slender, usually circular in cross section, glabrous except near nodes; bark of old stems rough and scaly; stem, in transverse section, shows pith appearing *cross shaped* because of four broad pith rays.

Buds solitary, oblong, pointed, with about three pairs of scales visible.

Leaf scars opposite, rather low, half elliptical; paired leaf scars connected by a transverse ridge. Bundle scar one, C-shaped, open side up. Stipule scars none.

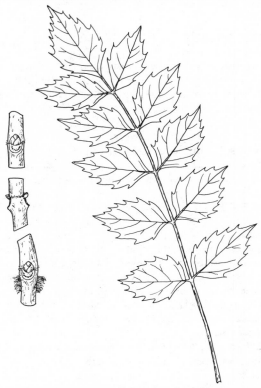

FIGURE 149   Trumpet creeper, *Campsis radicans*. Sections of stem showing lateral buds, leaf scars, and two bands of aerial rootlets below node; pinnately compound leaf.

Fruit a large, flat two-celled capsule, mostly 13–17 cm long, containing many flat, winged seeds.

Moist woods; infrequent at lower elevations, mostly below 2,000 feet, central and southwest NC; SC; GA; GSMNP.

Catalpa (catalpa)

Small or medium-sized trees with large, showy flowers in many-flowered clusters.

Leaves usually *three at a node* (whorled), sometimes two, deciduous, simple, *large,* broadly heart shaped, palmately veined, entire or occasionally angled along margin, long petioled.

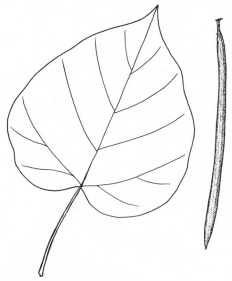

FIGURE 150   Southern catalpa, *Catalpa bignonioides*. Leaf and long, nearly cylindrical capsule.

Twigs stout, brittle, with prominent lenticels; pith *solid,* not chambered or hollow as in *Paulownia.*

Terminal bud absent; lateral buds solitary, small, appearing partly sunken in bark, with about six overlapping scales.

Leaf scars mostly three at a node, large, elliptical to nearly circular, each with raised margin and *depressed center.* Leaf scars at node not connected by transverse lines. Bundle scars numerous, arranged in an *ellipse or ring.* Stipule scars absent.

Flowers large, white, variously marked with yellow and purple, in clusters; May–June.

Fruit a *long, slender, podlike capsule,* usually persistent through winter, dangling from tree like a pencil; seeds numerous, flat, with wing bearded at each end.

1 Leaves abruptly short pointed, with unpleasant odor when crushed; bark of trunk thin and flaky; capsules usually less than 10 mm across at middle
.... (1) *C. bignonioides*
1' Leaves long pointed, with little or no detectable odor when crushed; bark of trunk thick and rough; capsules usually more than 10 mm across at middle
.... (2) *C. speciosa*

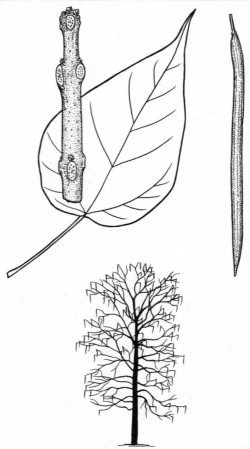

FIGURE 151   Northern catalpa, *Catalpa speciosa.* Leaf; section of twig showing buds and whorled leaf scars; cylindrical capsule; silhouette of mature tree bearing fruit.

| | 1. *Catalpa bignonioides* |
| FIGURE 150 | southern catalpa |

Mature bark *thin and scaly.*

Leaves whorled or opposite, 12–15 cm long, heart shaped, entire, *short pointed at apex,* light green and glabrous above, paler and hairy beneath; *giving off unpleasant odor when crushed.*

Twigs pale orange to brownish gray, minutely hairy.

Buds solitary, small, brownish.

Pods 15–40 cm long, more slender than those of C. *speciosa,* usually less than 10 mm thick at middle; seeds numerous, the *wings pointed* and with narrow fringe of hairs at end.

Native range uncertain, probably from western Florida to Louisiana; rarely escaped from cultivation in our area.

FIGURE 151 | **2. *Catalpa speciosa***
            | northern catalpa

Mature bark *thick, furrowed and ridged,* reddish brown to dark grayish brown.

Leaves usually whorled, sometimes opposite, 15–30 cm long, heart shaped, the margin entire or occasionally notched or lobed, *long pointed at apex,* dark green above at maturity, softly hairy beneath.

Twigs yellowish brown to greenish brown, glabrous.

Buds solitary, small and inconspicuous, brown.

Pods 20–50 cm long, *10–15 mm thick at the middle,* containing winged seeds, *the wings rounded* and hairy fringed.

Native range uncertain, probably from southern Illinois and Indiana to Arkansas and Tennessee; infrequently escaped from cultivation in our area.

Madder Family (Rubiaceae)

A large family of trees, shrubs, and herbs, mostly tropical; leaves opposite or whorled, simple, deciduous or evergreen, entire, with large stipules leaving ring scars at the nodes; flowers tube shaped, in clusters or globe-shaped heads; fruit a dry nutlet, capsule, or berry; buttonbush, partridgeberry.

| FIGURE 152 | *Cephalanthus occidentalis* buttonbush |

An upright, moisture-loving shrub.

Leaves *opposite or in whorls* of three or four, ovate to elliptical, 6–16 cm long, entire but often wavy margined, pointed at apex, broadly wedge shaped or somewhat rounded at base, dark green and glabrous above, lighter green and glabrous or hairy beneath.

Twigs moderately slender, rounded, reddish, or may be reddish on one side and green on the other, glabrous and glossy; lenticels raised, fairly prominent; pith light brown, more or less angled in cross section.

Buds *sunken in bark,* small, cone shaped, in a depression, often some distance above leaf scar; bud scales indistinct. Terminal bud absent.

Leaf scars opposite or whorled, raised, nearly circular; leaf scars at each node connected by a transverse stipular line or triangular stipules sometimes persistent. Bundle scar *one, crescent shaped or U-shaped.*

Flowers in *dense, ball-shaped heads,* 2–3 cm across; fruit a small dry nutlet clustered the same way.

Wet meadows, streambanks, low woods; infrequent, essentially throughout.

| | *Mitchella repens* partridgeberry |

A low, *creeping* plant, *only slightly woody,* commonly in large patches.

Leaves *opposite, evergreen,* simple, *circular-ovate,* 8–20 mm long and wide, entire rounded or heart shaped at base, glabrous, dark green and glossy on upper side, somewhat paler beneath, whitish along larger veins.

Stems slender, trailing, rooting at the nodes, short-hairy or glabrous.

Fruits *scarlet,* berrylike, *in united pairs,* 7–10 mm across, edible but tasteless, persistent all winter.

Rich, densely shaded woods; fairly common, throughout.

FIGURE 152   Buttonbush, *Cephalanthus occidentalis*. Leaf; section of twig showing whorled leaf scars; fruit.

Honeysuckle Family (Caprifoliaceae)

Shrubs, woody vines, trees, or herbs; leaves opposite, usually deciduous, simple or pinnately compound; flowers generally small, sometimes showy, often in clusters; fruit a capsule, berry, or drupe with one to several seeds; honeysuckle, viburnum, elder, and others.

Diervilla (bush-honeysuckle)

Low, soft-wooded shrubs.

Leaves opposite, deciduous, simple, toothed, short petioled or without petiole.

Twigs slender, marked with *longitudinal ridges,* two or more of them *hairy* and extending straight downward from lines connecting opposite leaf petioles or (in the absence of leaves) opposite leaf scars; bud scales persistent at base of young twigs.

Terminal bud present, except where replaced by terminal flower cluster or fruit cluster. Buds covered by four or more keeled, pointed scales.

Leaf scars only a little raised, V-shaped to crescent shaped. Opposing leaf scars meeting around stem or connected by a transverse ridge. Bundle scars three. Stipule scars none.

Fruit a capsule with *egg-shaped body* and *slender neck,* borne in terminal or subterminal clusters of two to six, persistent in winter.

1 Leaf petioles less than 4 mm long or lacking; twigs nearly square in cross
  section.. (1) *D. sessilifolia*
1' Leaf petioles 5–10 mm long; twigs nearly circular in cross section
  .... (2) *D. lonicera*

FIGURE 153 | 1. *Diervilla sessilifolia*
            | southern bush-honeysuckle

Leaves opposite, lanceolate, 5–18 cm long, finely toothed, with long, slender apex and rounded base, glabrous or nearly so on both surfaces; *petioles lacking or not more than 4 mm long.*

Twigs slender to moderate, *four-angled* in cross section, straw colored to brown or greenish brown, glabrous or nearly so; ridged downward from nodes, two (sometimes three or four) of the ridges hairy; bud scales persisting at base of young twigs.

Buds narrowly oblong, pointed, same color as twigs, with a few pairs of keeled, pointed scales.

Capsules 9–12 mm long, more or less bottle shaped.

Woods and balds; fairly common at higher elevations, mostly above 4,000 feet, central and southwest NC; SC; GSMNP.

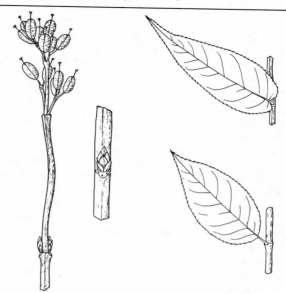

FIGURE 153   *Diervilla*. Left, center, and upper right: southern bush-honeysuckle, *D. sessilifolia*. Stem with fruit clusters, section of stem showing lateral bud and leaf scar, a single leaf with very short petiole. Lower right: northern bush-honeysuckle, *D. lonicera*. Leaf with longer petiole.

FIGURE 153	2. *Diervilla lonicera* northern bush-honeysuckle

Leaves opposite, lanceolate to ovate, 8–12 cm long, finely toothed, long pointed at apex, rounded at base, glabrous on both sides or thinly hairy along midrib beneath; *petioles 5–10 mm long*.

Twigs *nearly circular* in cross section but ridged downward from nodes, yellowish or brownish, glabrous.

Buds oblong, sharply pointed, appressed, with two or more pairs of scales.

Capsules 8–15 mm long.

Dry woods and rocky bluffs at higher elevations; Ashe, Buncombe, McDowell, Yancey Cos., NC.

Lonicera (honeysuckle)

Upright shrubs or twining vines.

Leaves *opposite*, deciduous or evergreen, simple, usually entire.

Twigs mostly slender; *bud scales persistent at base of current year's twigs*. Pith white or brown, solid or hollow.

Buds solitary or superposed (one above another), the lower one larger; bud scales two to many, opposite, four-ranked. Terminal bud present.

Leaf scars opposite, small, crescent shaped, raised on persistent base of leaf petioles. Bundle scars three. Stipule scars absent, but *opposing leaf scars connected by a transverse ridge.*

Fruit a *fleshy berry, with bitter taste.*

1 Stems erect, self-supporting:
  2 Pith of twigs brown, hollow between nodes . . . .(1) *L. bella*
  2' Pith of twigs white, solid between nodes . . . . (2) *L. canadensis*
1' Stems climbing or sprawling, not self-supporting:
  3 Upper leaves all separate, none joined together at base, and hence no opposing leaf scars forming a continuous ring around stem; stems usually hairy; leaves green beneath . . . . (3) *L. japonica*
  3' Upper leaves, especially those of lateral and flowering stems, joined together at base, making a double leaf, and hence each pair of upper leaf scars forming a continuous ring around stem; stems glabrous; leaves glaucous beneath:
    4 Leaves more or less evergreen and persistent; stems not glaucous; flowers and fruits in whorls spaced somewhat apart, forming a terminal spike. . . . .(4) *L. sempervirens*
    4' Leaves deciduous; stems glaucous; flowers and fruits in whorls crowded together, forming a dense terminal cluster . . . .(5) *L. dioica*

FIGURE 154 | **1. *Lonicera bella***
　　　　　　　　| belle honeysuckle

An erect shrub.

Leaves opposite, oblong, 3–5 cm long, entire, short pointed at apex but ending in a small abrupt tip (mucro), rounded at base; upper surface dull green and glabrous, lower surface slightly paler and hairy, or becoming nearly without hairs except on midrib and larger veins.

Twigs slender, ridged downward from nodes, yellowish brown or reddish brown to gray, more or less *hairy;* older twigs with *brown pith, the central part hollow.*

Buds often superposed, small, ovoid, same color as twigs, thinly hairy to glabrous.

Berries in pairs in axils of leaves, globose, red.

A garden hybrid, naturalized in a few places but rare in the mountains.

　　　　　　　| **2. *Lonicera canadensis***
　　　　　　　| fly honeysuckle

An erect shrub.

Leaves opposite, all separate and having petioles, rather thin, oblong-ovate, 4–9 cm long, entire, *fringed with hairs along margin,* blunt or short pointed at apex, rounded or heart shaped at base, bright green and *glabrous* on both surfaces.

FIGURE 154   Belle honeysuckle, *Lonicera bella*. Stem with leaves and fruit; section of stem showing superposed buds.

Twigs slender, flexuous, ridged downward from nodes, ashy gray, glabrous; *pith white, solid.*

Buds divergent, conical to ovoid, 3–6 mm long, the scales close fitting, short pointed.

Berries in pairs in axils of leaves, ellipsoid to globose, 8–10 mm long, distinct or slightly joined at base, appearing as twin berries, red when ripe.

Moist rocky woods at higher elevations, mostly above 4,000 feet; Swain, Watauga, and Yancey Cos., NC; GA; GSMNP.

FIGURE 155   **3. *Lonicera japonica***
Japanese honeysuckle

A climbing or sprawling vine, often forming *dense tangles.*

Leaves opposite, evergreen or mostly so, some turning reddish or purplish in winter, all leaves separate, *no pairs united,* ovate to oblong, 4–8 cm long, usually entire but sometimes lobed on young stems in spring, fringed with hairs along margin, pointed at apex, rounded at base, glossy dark green above, green or purplish and hairy to glabrous beneath.

Stems *twining,* light reddish brown, usually densely hairy, bark soon peeling and showing lighter-colored underbark; *stems hollow (pith lacking).*

Buds solitary, 1–4 mm long, reddish brown to greenish brown, the scales loosely fitting, hairy; stems commonly overwinter without buds, but with clusters of small leaves in axils of larger ones.

Berries solitary or in pairs along stems, in axils of leafy bracts, globose, 5–8 mm across, smooth, *black,* glossy.

Native of Asia, escaped from cultivation and now abundant in some places.

FIGURE 155   **4. *Lonicera sempervirens***
coral honeysuckle

A climbing or trailing vine. Older bark yellowish brown, shredding into long, fibrous strips.

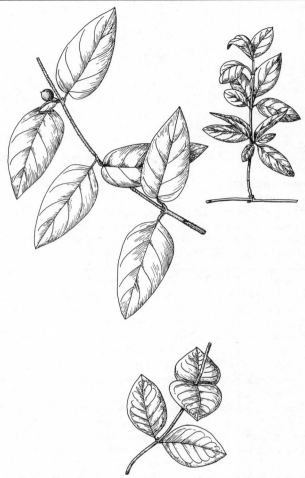

FIGURE 155   *Lonicera*. Above: Japanese honeysuckle, *L. japonica*. Left: Stem, leaves, and fruit. Right: Stem and leaves, with clusters of small leaves in axils of larger ones. Below: coral honeysuckle, *L. sempervirens*. Stem and leaves, the upper pair of leaves united at base.

Leaves opposite, more or less evergreen, elliptical to ovate, 3–7 cm long, entire but *the uppermost pairs, just below the flowers, united at base;* usually short pointed at apex, dark green above, *glaucous beneath,* glabrous on both sides.

Stems *twining,* slender, yellowish brown or yellowish green, glabrous; bark shredding in second year. Pith *pale or brown, hollow* between nodes, with only a thin layer around hollow center.

Buds solitary, ovoid, 3–6 mm long, brown, the scales keeled, pointed, spreading at tip.

Leaf scars narrow, extending about halfway around stem, expanded at center.

Flowers *long, slender,* trumpet shaped, coral colored.

Berries in clusters of three to six near the end of stems, ovoid, 8–10 mm long, 6–7 mm thick, bright red.

Woods, thickets, fencerows; Henderson Co., NC; Greenville, Oconee, Pickens Cos., SC; GA; GSMNP.

5. *Lonicera dioica*
mountain honeysuckle

A trailing or twining, or sometimes bushy, shrub.

Leaves opposite, deciduous, 6–15 cm long, variable in shape, the uppermost one to four pairs *united at base* to form double leaves, entire, short pointed with mucro or bristlelike awn at apex, glabrous on both surfaces, very glaucous beneath.

Stems gray or straw colored, glabrous, *glaucous.*

Flowers and fruits several in a *compact* cluster at ends of stems.

Fruit ellipsoid to globose, 7–10 mm long, red to salmon colored.

Dry or moist woods and thickets; Ashe, Buncombe, Macon, Watauga, Yancey Cos., NC.

Symphoricarpos orbiculatus
FIGURE 156 | coralberry, Indian-currant

A small shrub with very slender branches.

Leaves *opposite,* deciduous, simple, broadly elliptical to ovate, 2–4 cm long, entire or wavy margined, blunt or rounded at apex, wedge shaped or rounded at base, dull green above, paler and hairy beneath.

Twigs *very slender,* purplish or brownish, more or less hairy; *older bark shreddy;* bud scales persisting where twigs of the current season join twigs of the previous year.

Buds small, 2–3 mm long, oblong to ovoid, somewhat flattened, with about three pairs of keeled, pointed scales showing. Terminal bud lacking.

Leaf scars opposite, small, raised, half circular; each pair of leaf scars connected around twig by a raised line or ridge. Bundle scar *one,* dotlike or indistinct. Stipule scars none.

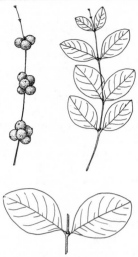

FIGURE 156   Coralberry, *Symphoricarpos orbiculatus.* Twig with fruit clusters, twig with leaves, section of twig with two leaves.

Fruit a globose, *purplish red,* two-seeded berry, 5–8 mm across, crowded in small axillary clusters, persisting well into winter.

Old fields, pastures, roadsides, woods along streams; infrequent, essentially throughout.

Viburnum (viburnum)

Shrubs or occasionally small trees.

Leaves *opposite,* deciduous, simple, either pinnately or palmately veined, mostly toothed or lobed, or both, seldom nearly entire.

Twigs slender to stout, typically *dull colored,* not highly colored as in *Cornus* and some species of *Acer.*

Buds naked, or covered by a single pair of valvate scales, or with two or three pairs of overlapping scales. Terminal bud present, except on those twigs having a terminal fruit scar. Buds commonly *of two kinds:* large, swollen flower buds and fairly slender leaf buds.

Leaf scars opposite, usually narrow, crescent shaped to V-shaped or triangular, constricted between bundle scars, or occasionally large and three-lobed; *opposing leaf scars meeting around stem or connected by a transverse ridge.* Bundle scars three. Stipule scars lacking.

Fruit a *one-seeded drupe,* borne in terminal clusters, ripening in summer or autumn.

Members of this genus may be confused with members of *Cornus* because the leaves are opposite and simple, but they are almost always toothed or lobed (usually entire only in *V. nudum* and occasionally entire in *V. cassinoides*), whereas in the dogwoods the leaves are entire.

Summer Key

1 Leaves mostly three-lobed, palmately veined, rarely some leaves unlobed or the lobes shallow or obscure . . . . (1) *V. acerifolium*

1′ Leaves not lobed, pinnately veined:

  2 Leaf blades mostly 10 cm wide or more; buds, young twigs, and often the lower surface of leaves covered with a scurfy layer of star-shaped hairs . . . . (2) *V. alnifolium*

  2′ Leaf blades less than 10 cm wide:

    3 Leaves coarsely toothed, all or most lateral veins ending in a tooth:

      4 Petioles and lower surface of leaves hairy . . . . (3) *V. dentatum*

      4′ Petioles glabrous; lower surface of leaves glabrous or hairy only on veins or in vein axils . . . . (4) *V. dentatum* var. *lucidum*

    3′ Leaves finely toothed to entire, the lateral veins merging into a fine pattern of veinlets before reaching leaf margin:

      5 Margin of leaves distinctly and regularly toothed, the teeth pointing forward (toward leaf tip):

        6 Leaves rather thick, their upper surface glossy; petioles rusty-woolly and with a broad, wavy margin . . . . (5) *V. rufidulum*

        6′ Leaves thin, their upper surface dull green; petioles not rusty-woolly and without broad, wavy margin . . . . (6) *V. prunifolium*

      5′ Margin of leaves bluntly or irregularly toothed to entire; teeth, if present, pointing outward:

        7 Leaves all or mostly toothed, the teeth more or less rounded . . . . (7) *V. cassinoides*

        7′ Leaves entire, or mostly entire but some leaves wavy margined or obscurely toothed:

          8 Leaves glossy above; lower surface minutely hairy near and along margin; flower and fruit clusters on a long basal stalk, just above the topmost pair of leaves, the stalk usually longer than the several branches of the cluster . . . . (8) *V. nudum*

          8′ Leaves dull above; lower surface without hairs near and along margin; flower and fruit clusters on a short basal stalk, just above the topmost pair of leaves, the stalk usually shorter than the several branches of the cluster . . . . (7) *V. cassinoides*

Winter Key

1 Buds naked (without true scales), some or all showing fleshy, folded, rudimentary leaves. . . . . (2) *V. alnifolium*

1′ Buds with valvate or overlapping scales:

  2 Buds with one pair of valvate scales, both scales as long as the bud:

    3 Buds covered with dark red woolly hairs; young twigs more or less rusty-woolly . . . . (5) *V. rufidulum*

    3′ Buds not covered with dark red woolly hairs:

      4 Lateral twigs stiff and usually short, set at wide angle to central stem; buds grayish brown or brownish pink to lead gray, usually somewhat glossy. . . . . (6) *V. prunifolium*

      4′ Lateral twigs flexible and usually longer, ascending at a narrower angle to central stem; buds not at all glossy:

        5 Terminal bud reddish brown; fruit remnants on a basal stalk 1–5 cm long, just above the uppermost pair of leaf scars, the stalk usually as long as or longer than the branches of the cluster . . . . (8) *V. nudum*

        5′ Terminal bud yellowish brown or golden; fruit remnants on a basal stalk 5–17 mm long, just above the uppermost pair of leaf scars, shorter than the branches of the cluster . . . . . . (7) *V. cassinoides*

  2′ Buds with more than one pair of scales, the outer pair shorter than the bud:

    6 Buds green or greenish purple; lateral buds with the lowest pair of scales very short, one-fourth to one-third as long as the bud . . . . (1) *V. acerifolium*

    6′ Buds brown or only slightly tinged with green; lateral buds with the lowest pair of scales usually one-half or more as long as the bud:

      7 Twigs and outer bud scales hairy. . . . .(3) *V. dentatum*

      7′ Twigs and outer bud scales glabrous or with only scattered hairs . . . .(4) *V. dentatum* var. *lucidum*

FIGURE 157 | 1. *Viburnum acerifolium*
mapleleaf viburnum

An erect shrub, usually in colonies.

Leaves 7–13 cm long and about as wide, *three-lobed* or mostly so, palmately veined, coarsely toothed, much like leaves of red maple (*Acer rubrum*), heart shaped to straight across at base, finely hairy to nearly glabrous and *red dotted or black dotted* on lower surface (magnifying lens needed).

Twigs slender, six-sided, grayish brown, hairy or glabrous.

Terminal bud 5–10 mm long, *green or greenish purple,* glabrous or essentially so, slender and pointed if a leaf bud, swollen near base if a flower bud, both kinds with two pairs of scales visible, the lower pair very short. Lateral buds 4–7 mm long, appressed, short stalked, slender and pointed, same color as terminal bud, with four scales showing, the first pair usually one-third or less as long as bud.

Fruit 8–9 mm long, dark blue or bluish black, without bloom.

Moist or dry woods; fairly common, throughout.

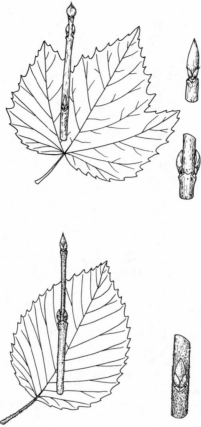

FIGURE 157   *Viburnum.* Above: mapleleaf viburnum, *V. acerifolium.* Leaf; twig ending in terminal flower bud; detailed sections of twig showing terminal leaf bud and two lateral leaf buds. Below: southern arrowwood, *V. dentatum.* Leaf, twig with terminal bud and a pair of lateral buds; detailed section of twig showing lateral bud and leaf scar.

| FIGURE 158 | 2. *Viburnum alnifolium* |
| | hobblebush, moosewood |

An irregularly branched shrub, the lower branches sometimes bending over and rooting near tip.

Leaves *large,* broadly ovate to nearly circular, 10–25 cm long, 8–20 cm wide, pinnately veined, finely double toothed, mostly short pointed at apex, heart shaped

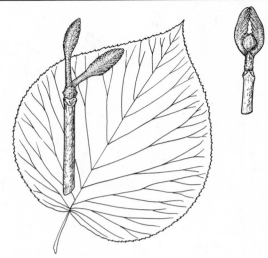

FIGURE 158   Hobblebush, *Viburnum alnifolium*. Leaf; twig showing a pair of leaf buds at end of twig; a flower bud.

to rounded at base, glabrous or practically so above, densely to sparingly scurfy-hairy beneath, especially on veins.

Twigs stout to moderately slender, purple to olive brown, rusty-scurfy to nearly glabrous. Branching sometimes whorled.

Terminal bud *large, naked, stalked,* reddish brown, *densely scurfy-hairy.* Leaf buds long stalked, mostly 15–30 mm long, including the stalk; flower buds rounded, flanked by a pair of leaf buds.

Leaf scars large, broad, more or less three-lobed, widely spaced along twigs, occasionally whorled.

Fruits ovoid, 8–12 mm long, bright red, eventually turning black.

Moist woods, cool ravines, edges of streams at high elevations; common above 4,500 feet.

	3. *Viburnum dentatum*
FIGURE 157	southern arrowwood

An erect shrub.

Leaves broadly ovate to elliptical-ovate or lanceolate-ovate, 5–12 cm long, 4–10 cm wide, *coarsely toothed* with sharp, outward-pointing teeth, long pointed or short pointed at apex, rounded or broadly wedge shaped at base, slightly hairy above, *covered with tufted hairs beneath;* lateral veins prominent, straight or nearly so, simple or forked, each vein extending to the margin and ending in a tooth; *petioles densely hairy.*

Twigs angled or six-sided, ridged below leaf scars, grayish brown to reddish brown, *hairy with clustered hairs, rough to touch.*

Buds reddish brown to greenish, with two or three pairs of scales exposed, the scales keeled, more or less hairy. Terminal bud larger than laterals. Lateral buds appressed, the tips curved inward toward twig; bud scales two pairs, ciliate, strongly keeled, the outer pair nearly one-half to more than one-half as long as the bud.

Fruit bluish to black, 6–10 mm long, often hairy.

Thickets, open woods; occasional, southwest mountains of NC and southward; GSMNP.

4. *Viburnum dentatum* var. *lucidum*
northern arrowwood

Leaves similar to those of *V. dentatum* but nearly or quite glabrous; lower surface may have tufts of hairs in vein axils or a few scattered hairs along veins. Twigs smooth and glabrous or essentially so. Buds glabrous.

Low or wet woods; Buncombe, Henderson, and Macon Cos., NC; GA; GSMNP.

5. *Viburnum rufidulum*
rusty blackhaw

A large shrub or small tree. Bark of large trunks nearly black, deeply checkered into squarish blocks.

Leaves thick and somewhat leathery, elliptical to ovate, 5–10 cm long, short pointed, blunt, or notched at apex, rounded or broadly wedge shaped at base, *glossy green above,* covered with *rust-colored hairs on lower side,* at least along the veins; petioles hairy, with broad, wavy margin.

Twigs slender, densely rusty-hairy when young, changing to gray and nearly glabrous; lateral twigs short, somewhat more flexible than those of *V. prunifolium.*

Terminal bud more or less flattened, blunt, *rusty red–woolly* or scurfy, with two valvate scales. Flower buds 9–12 mm long, abruptly swollen below middle; leaf buds smaller, 5–7 mm long.

Fruit small, 12–15 mm long, blue with whitish bloom; flesh slightly sweet, edible.

Mostly in dry woods; Henderson and McDowell Cos., NC; Oconee Co., SC; GSMNP.

6. *Viburnum prunifolium*
FIGURE 159 | blackhaw

A large shrub or small tree. Bark of trunk reddish brown to blackish, made up of *small, thick, squarish scales,* similar to bark of *Cornus florida.*

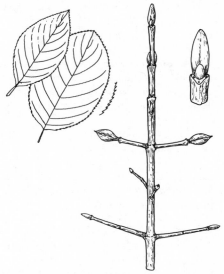

FIGURE 159　　Blackhaw, *Viburnum prunifolium*. Two forms of leaves, twig showing leaf
buds, terminal flower buds, and wide-angle branching; enlarged terminal
leaf bud, uppermost lateral bud, and leaf scar.

Leaves elliptical to elliptical-obovate or occasionally nearly circular, 2–10 cm long,
very finely toothed with incurved teeth, blunt to rather long pointed at apex,
rounded or wedge shaped at base, glabrous or nearly so on both surfaces, or
sometimes scurfy on upper midrib.

Twigs slender, ridged downward from nodes, reddish brown to yellowish brown,
frequently with grayish skin. Lateral twigs, or most of them, *short and stiff,* often
spreading at nearly right angle to main stem.

Buds pinkish to ashy brown or gray, with *valvate scales,* the scales appearing rather
smooth to the naked eye but under magnifying lens seen to be roughened by
waxy bloom or flakes. Flower buds terminal, 9–12 mm long, swollen near
middle, tapering above to a rounded tip; terminal leaf buds more slender, 5–12
mm long. Lateral buds 3–7 mm long, narrowly oblong, bluntly pointed, more
or less appressed.

Flower and fruit clusters *without a stalk.*

Fruit blue to black, 9–14 mm long; flesh with sweet, pleasant taste when ripe.

Woods, thickets, roadsides; fairly common, essentially throughout, but rare in
GSMNP.

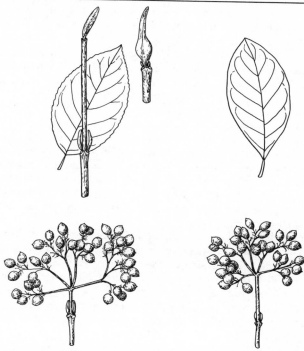

FIGURE 160   *Viburnum*. Left: withe-rod, *V. cassinoides*. Leaf; section of stem showing terminal leaf bud and a pair of lateral buds; end of twig bearing a terminal flower bud; terminal fruit cluster. Right: possumhaw, *V. nudum*. Leaf, terminal fruit cluster.

FIGURE 160 | **7. *Viburnum cassinoides***
withe-rod

A large shrub.

Leaves ovate to lanceolate, 5–11 cm long, distinctly toothed with rather *blunt teeth,* or occasionally nearly entire, the margin usually not rolled under (revolute); blunt to long pointed at apex, wedge shaped to rounded at base; lateral veins not clearly extending to the teeth; *dull* green above, *hairless* or somewhat scurfy and reddish dotted beneath.

Twigs slender, angled below nodes, medium brown to grayish brown, rusty-scurfy when young and often remaining so, especially near tip.

Buds golden brown, scurfy, appearing pitted (under magnifying lens), with two *valvate scales*. Flower buds terminal, 15–25 mm long or occasionally longer, swollen near base; terminal leaf buds slender, 10–20 mm long. Lateral buds slender and often curved, loosely appressed, 6–15 mm long.

Flower and fruit clusters *short stalked.*

Fruits 6–8 mm long, blue or black when mature, with heavy bloom; flesh sweet.

Moist upland woods; fairly common, throughout, mostly above 4,500 feet.

---

	8. *Viburnum nudum*
FIGURE 160	possumhaw

A medium-sized to large shrub or (rarely) a small tree.

Leaves mostly oblong-lanceolate, 5–12 cm long, *entire,* or mostly entire but some leaves wavy margined or obscurely toothed, the teeth pointing outward; margin commonly narrowly revolute; principal veins not clearly extending to leaf margin; short pointed or long pointed at apex, wedge shaped or rounded at base; upper surface *glossy green,* lower surface thinly hairy, at least along margin, and covered with rusty dots.

Twigs slender, slightly scurfy, reddish brown, more or less glossy.

Buds reddish brown to grayish brown, scurfy, with *valvate scales.* Terminal flower buds 12–20 mm long, enlarged at base, long pointed; terminal leaf buds about as in *V. cassinoides.* Lateral buds slender, appressed, mostly 6 mm long or less.

Flower and fruit clusters *long stalked.*

Fruit about 8 mm long, blue black when mature; flesh bitter.

Low woods, boggy places, edges of streams; occasional, at lower elevations; NC, SC, GA.

Sambucus (elder)

Shrubs with *soft wooded, pithy stems.*

Leaves *opposite,* deciduous, *pinnately compound,* the leaflets odd numbered and toothed.

Twigs stout, many angled in cross section; lenticels prominent, raised, corky. Pith large and soft.

Terminal bud absent. Lateral buds solitary or multiple (two or more together), each enclosed by two to five pairs of scales; twigs ending with uppermost pair of lateral buds or with fruit scar.

Leaf scars opposite, *large;* opposing leaf scars meeting around stem or connected by a transverse line. Bundle scars three to seven, usually five. Stipules and stipule scars absent.

Fruit *berrylike,* blackish or red, *juicy,* in terminal clusters, ripe in summer.

1 Pith white; buds broad based, small, mostly 2–5 mm long; flower and fruit clusters flat topped or convex ....(1) *S. canadensis*

1′ Pith orange or brown; buds narrowed at base, large, mostly more than 7 mm long; flower and fruit clusters pyramid shaped...................(2) *S. pubens*

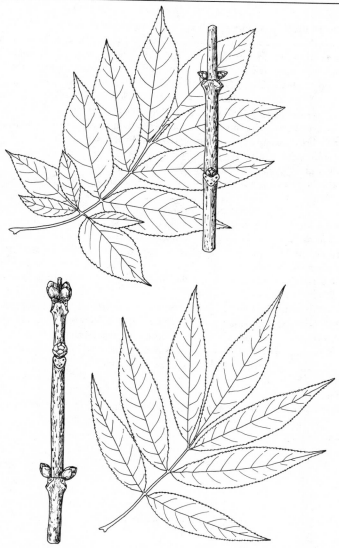

FIGURE 161   *Sambucus*. Above: American elder, *S. canadensis*. Pinnately compound leaf; section of twig showing lateral buds, leaf scars, and bundle scars. Below: red-berried elder, *S. pubens*. Detailed section of twig showing dying back of twig apex, uppermost pair of lateral buds, lower lateral buds, and leaf scars; pinnately compound leaf.

FIGURE 161	1. *Sambucus canadensis* American elder, common elderberry

Leaflets usually seven to nine (five to eleven), lanceolate to ovate, 5–15 cm long, the lowest pair often divided into two to four subleaflets; sharply toothed, long pointed at apex, glabrous above, often hairy beneath, at least along veins.

Twigs mostly stout, sparingly branched, soft wooded, easily broken, more or less angled, yellowish gray, pinkish gray, or greenish gray, glabrous. Pith *white*.

Buds solitary or multiple, small, less than 5 mm long, *conical* or ovoid, divergent, reddish brown to green, glabrous, with several pairs of scales visible.

Flowers and fruits in *flat topped* clusters; fruit clusters heavy, drooping; fruit *shiny black* or dark purple, 4–6 mm across, juicy and slightly sweet, edible.

Open woods, clearings, roadsides, streambanks; common, throughout.

FIGURE 161	2. *Sambucus pubens* red-berried elder, red elderberry

Leaflets five to seven, ovate-lanceolate to lanceolate, 6–15 cm long, finely and sharply toothed, bright green and glabrous on upper side, hairy or glabrous beneath; terminal leaflet long stalked, lateral leaflets short stalked to nearly stalkless.

Twigs stout, stiff, obscurely many angled near tip, brownish green or yellowish green to greenish gray, hairy to glabrous; lenticels prominent, orange colored. Pith large, *brown* to orange. Bark has *rank odor* when bruised.

Buds solitary or multiple, the larger buds almost *globose*, 7–12 mm long, somewhat stalked, greenish or reddish, with several pairs of scales visible.

Flowers and fruits in *cone-shaped* clusters; fruits *red,* 3–5 mm across, sour, not edible.

Rich woods, mostly at higher elevations, throughout, common above 4,000 feet.

List of Trees and Shrubs Arranged according to Family

Gymnosperms
Pine Family (Pinaceae)

Pinus L. (pine)
  P. echinata Mill.—shortleaf pine
  P. pungens Lamb.—Table Mountain pine
  P. rigida Mill.—pitch pine
  P. strobus L.—eastern white pine
  P. virginiana Mill.—Virginia pine
Picea A. Dietr. (spruce)
  P. rubens Sarg.—red spruce
Tsuga (Endl.) Carr. (hemlock)
  T. canadensis (L.) Carr.—eastern hemlock
  T. caroliniana Engelm.—Carolina hemlock
Abies Mill. (fir)
  A. fraseri (Pursh) Poir.—Fraser fir

Cypress Family (Cupressaceae)
Thuja L. (thuja)
  T. occidentalis L.—northern white-cedar
Juniperus L. (juniper)
  J. virginiana L.—eastern redcedar

Angiosperms: Monocotyledons
Lily Family (Liliaceae)
Smilax L. (greenbrier, catbrier)
  S. bona-nox L.—sawbrier
  S. glauca Walt.—glaucous greenbrier
  S. hispida Muhl.—bristly greenbrier
  S. rotundifolia L.—common greenbrier

Angiosperms: Dicotyledons
Willow Family (Salicaceae)
Salix L. (willow)
  S. alba L.—white willow
  S. babylonica L.—weeping willow
  S. cinerea L.—gray willow
  S. humilis Marsh.—upland willow
  S. nigra Marsh.—black willow
  S. sericea Marsh.—silky willow

Populus L. (cottonwood, poplar)
  P. alba L.—white poplar
  P. deltoides Bartr. ex. Marsh.—eastern cottonwood
  P. gileadensis Roleau—balm-of-Gilead poplar
  P. grandidentata Michx.—bigtooth aspen
  P. nigra var. *italica* Muenchh.—Lombardy poplar

Sweet Gale Family (Myricaceae)
  Comptonia L'Her. (sweet-fern)
    C. peregrina (L.) Coult.—sweet-fern

Walnut Family (Juglandaceae)
  Juglans L. (walnut)
    J. cinerea L.—butternut
    J. nigra L.—black walnut
  Carya Nutt. (hickory)
    C. cordiformis (Wangenh.) K. Koch—bitternut hickory
    C. glabra (Mill.) Sweet—pignut hickory
    C. ovata (Mill.) K. Koch—shagbark hickory
    C. pallida (Ashe) Engl. & Graebn.—sand hickory
    C. tomentosa (Poir.) Nutt.—mockernut hickory

Birch Family (Betulaceae)
  Betula L. (birch)
    B. alleghaniensis Britton—yellow birch
    B. lenta L.—sweet birch
    B. nigra L.—river birch
    B. papyrifera Marsh.—paper birch
    B. populifolia Marsh.—gray birch
  Alnus Mill. (alder)
    A. crispa (Ait.) Pursh—mountain alder
    A. serrulata (Ait.) Willd.—hazel alder
  Carpinus L. (hornbeam)
    C. caroliniana Walt.—American hornbeam
  Ostrya Scop. (hophornbeam)
    O. virginiana (Mill.) K. Koch—eastern hophornbeam
  Corylus L. (hazelnut)
    C. americana Walt.—American hazelnut
    C. cornuta Marsh.—beaked hazelnut

Beech Family (Fagaceae)
  Fagus L. (beech)
    F. grandifolia Ehrh.—American beech

Castanea Mill. (chestnut, chinkapin)
    C. dentata (Marsh.) Borkh.—American chestnut
    C. pumila Mill.—Allegheny chinkapin
Quercus L. (oak)
    Q. alba L.—white oak
    Q. coccinea Muenchh.—scarlet oak
    Q. falcata Michx.—southern red oak
    Q. imbricaria Michx.—shingle oak
    Q. marilandica Muenchh.—blackjack oak
    Q. meuhlenbergii Engelm.—chinkapin oak
    Q. prinoides Willd.—dwarf chinkapin oak
    Q. prinus L.—chestnut oak
    Q. rubra L.—northern red oak
    Q. stellata Wangenh.—post oak
    Q. velutina Lam.—black oak

Elm Family (Ulmaceae)

Ulmus L. (elm)
    U. alata Michx.—winged elm
    U. americana L.—American elm
    U. rubra Muhl.—slippery elm
Celtis L. (hackberry)
    C. laevigata Willd.—sugarberry
    C. occidentalis L.—hackberry

Mulberry Family (Moraceae)

Maclura Nutt. (Osage-orange)
    M. pomifera (Raf.) Schneid.—Osage-orange
Broussonetia L'Her. (paper-mulberry)
    B. papyrifera (L.) Vent.—paper-mulberry
Morus L. (mulberry)
    M. alba L.—white mulberry
    M. rubra L.—red mulberry

Sandalwood Family (Santalaceae)

Pyrularia Michx. (buffalo nut)
    P. pubera Michx.—buffalo nut

Mistletoe Family (Loranthaceae)

Phoradendron Nutt. (mistletoe)
    P. flavescens (Pursh) Nutt.—American mistletoe

Birthwort Family (Aristolochiaceae)
    Aristolochia L. (Dutchman's-pipe)
        A. durior Hill.—Dutchman's-pipe

Buttercup Family (Ranunculaceae)
    Xanthorhiza Marsh. (yellowroot)
        X. simplicissima Marsh.—yellowroot
    Clematis L. (clematis)
        C. viorna L.—leather-flower
        C. virginiana L.—virgin's bower

Barberry Family (Berberidaceae)
    Berberis L. (barberry)
        B. canadensis Mill.—American barberry
        B. thunbergii DC.—Japanese barberry

Moonseed Family (Menispermaceae)
    Cocculus DC. (snailseed)
        C. carolinus (L.) DC.—coralbeads
    Menispermum L. (moonseed)
        M. canadense L.—moonseed

Magnolia Family (Magnoliaceae)
    Magnolia L. (magnolia)
        M. acuminata L.—cucumbertree
        M. fraseri Walt.—Fraser magnolia
        M. tripetala L.—umbrella magnolia
    Liriodendron L. (yellow-poplar)
        L. tulipifera L.—yellow-poplar

Calycanthus Family (Calycanthaceae)
    Calycanthus L. (sweetshrub)
        C. fertilis Walt.—sweetshrub
        C. floridus L.—hairy sweetshrub

Custard-apple Family (Annonaceae)
    Asimina Adans. (pawpaw)
        A. triloba (L.) Dunal—pawpaw

Laurel Family (Lauraceae)
    Lindera Thunb. (spicebush)
        L. benzoin (L.) Bl.—spicebush
    Sassafras Nees (sassafras)
        S. albidum (Nutt.) Nees—sassafras

Saxifrage Family (Saxifragaceae)
Itea L. (sweet-spires)
    I. virginica L.—sweet-spires
Decumaria L. (decumaria)
    D. barbara L.—southern decumaria
Philadelphus L. (mock-orange)
    P. hirsutus Nutt.—hairy mock-orange
    P. inodorus L.—scentless mock-orange
Hydrangea L. (hydrangea)
    H. arborescens L.—wild hydrangea
Ribes L. (gooseberry, currant)
    R. cynosbati L.—pasture gooseberry
    R. glandulosum Grauer—skunk currant
    R. rotundifolium Michx.—roundleaf gooseberry

Witch-hazel Family (Hamamelidaceae)
Hamamelis L. (witch-hazel)
    H. virginiana L.—witch-hazel
Liquidambar L. (sweetgum)
    L. styraciflua L.—sweetgum
Fothergilla Murr. (fothergilla)
    F. major (Sims) Lodd.—large fothergilla

Plane-tree Family (Platanaceae)
Platanus L. (sycamore)
    P. occidentalis L.—sycamore

Rose Family (Rosaceae)
Physocarpus Maxim. (ninebark)
    P. opulifolius (L.) Maxim—ninebark
Spiraea L. (spiraea)
    S. alba Du Roi—narrowleaf spiraea
    S. japonica L. f.—Japanese spiraea
    S. latifolia Borkh.—broadleaf spiraea
    S. tomentosa L.—steeplebush spiraea
    S. virginiana Britton—Virginia spiraea
Aronia Medic. (chokeberry)
    A. arbutifolia (L.) Ell.—red chokeberry
    A. melanocarpa (Michx.) Ell.—black chokeberry
Crataegus L. (hawthorn)
    C. crus-galli L.—cockspur hawthorn
    C. punctata Jacq.—dotted hawthorn
Sorbus L. (mountain-ash)
    S. americana Marsh.—American mountain ash

Amelanchier Medic. (serviceberry)
    A. arborea (Michx. f.) Fern.—downy serviceberry
    A. sanguinea (Pursh) DC.—roundleaf serviceberry
Malus Mill. (apple)
    M. angustifolia (Ait.) Michx.—southern crab apple
    M. coronaria (L.) Mill.—sweet crab apple
    M. sylvestris (L.) Mill.—apple
Pyrus L. (pear)
    P. communis L.—pear
Rubus L. (blackberry, dewberry, raspberry)
    R. allegheniensis Porter—Allegany blackberry
    R. argutus Link—tall blackberry
    R. canadensis L.—smooth blackberry
    R. flagellaris Willd.—dewberry
    R. hispidus L.—swamp dewberry
    R. idaeus var. *canadensis* Richardson—red raspberry
    R. occidentalis L.—black raspberry
    R. odoratus L.—flowering raspberry
    R. phoenicolasius Maxim.—wineberry
Rosa L. (rose)
    R. carolina L.—Carolina rose
    R. micrantha Smith—sweetbrier
    R. multiflora Thunb.—multiflora rose
    R. palustris Marsh.—swamp rose
    R. setigera Michx.—climbing rose
Prunus L. (cherry, plum, peach)
    P. americana Marsh.—American plum
    P. angustifolia Marsh.—Chickasaw plum
    P. avium (L.) L.—mazzard
    P. cerasus L.—sour cherry
    P. pensylvanica L.—pin cherry
    P. persica Batsch—peach
    P. serotina Ehrh.—black cherry
    P. virginiana L.—chokecherry

Legume Family (Leguminosae)
Albizia Durazzini (albizia)
    A. julibrissin Durazzini—silktree
Cercis L. (redbud)
    C. canadensis L.—redbud
Gleditsia L. (honeylocust)
    G. triacanthos L.—honeylocust
Gymnocladus Lam. (coffeetree)
    G. dioicus (L.) K. Koch—Kentucky coffeetree

Cladrastis Raf. (yellowwood)
 C. kentukea (Dum.-Cours.) Rudd—yellowwood
Cytisis L. (broom)
 C. scoparius (L.) Link—Scotch broom
Amorpha L. (indigobush)
 A. fruticosa L.—false-indigo
 A. glabra Poiret—mountain-indigo
Wisteria Nutt. (wisteria)
 W. sinensis (Sims) Sweet—Chinese wisteria
Robinia L. (locust)
 R. boyntonii Ashe—Boynton locust
 R. hispida L.—bristly locust
 R. kelseyi Hutch.—Kelsey locust
 R. pseudoacacia L.—black locust
 R. viscosa Vent.—clammy locust

Rue Family (Rutaceae)
Zanthoxylum L. (prickly-ash)
 Z. americanum Mill.—prickly-ash
Ptelea L. (hoptree)
 P. trifoliata L.—hoptree

Quassia Family (Simaroubaceae)
Ailanthus Desf. (ailanthus)
 A. altissima (Mill.) Swingle—ailanthus

Mahogany Family (Meliaceae)
Melia L. (chinaberry)
 M. azedarach L.—chinaberry

Cashew Family (Anacardiaceae)
Rhus L. (sumac)
 R. aromatica Aiton—fragrant sumac
 R. copallina L.—shining sumac
 R. glabra L.—smooth sumac
 R. typhina L.—staghorn sumac
Toxicodendron Mill. (poison-ivy, poison-sumac)
 T. radicans (L.) Kuntze—poison-ivy
 T. vernix (L.) Kuntze—poison-sumac

Holly Family (Aquifoliaceae)
Ilex L. (holly)
 I. montana Torr. & Gray—mountain winterberry
 I. opaca Ait.—American holly
 I. verticillata (L.) Gray—common winterberry

Staff-tree Family (Celastraceae)

Euonymus L. (euonymus)

    E. americanus L.—strawberry-bush

    E. atropurpureus Jacq.—burningbush

    E. obovatus Nutt.—running strawberry-bush

Celastrus L. (bittersweet)

    C. orbiculatus Thunb.—Oriental bittersweet

    C. scandens L.—American bittersweet

Bladdernut Family (Staphylaceae)

Staphylea L. (bladdernut)

    S. trifolia L.—bladdernut

Maple Family (Aceraceae)

Acer L. (maple)

    A. negundo L.—boxelder

    A. nigrum Michx. f.—black maple

    A. pensylvanicum L—striped maple

    A. rubrum L.—red maple

    A. saccharinum L.—silver maple

    A. saccharum Marsh.—sugar maple

    A. spicatum Lam.—mountain maple

Horse-chestnut Family (Hippocastanaceae)

Aesculus L. (buckeye)

    A. octandra Marsh.—yellow buckeye

    A. sylvatica Bartr.—painted buckeye

Buckthorn Family (Rhamnaceae)

Rhamnus L. (buckthorn)

    R. caroliniana Walt.—Carolina buckthorn

    R. frangula L.—glossy buckthorn

Ceanothus L. (ceanothus)

    C. americanus L.—New Jersey tea

Grape Family (Vitaceae)

Vitis L. (grape)

    V. aestivalis Michx.—summer grape

    V. aestivalis var. *argentifolia* (Munson) Fern.—silver-leaf grape

    V. baileyana Munson—possum grape

    V. cinerea Englem.—sweet winter grape

    V. labrusca L.—fox grape

    V. rotundifolia Michx.—muscadine

    V. vulpina L.—frost grape

Ampelopsis Michx. (ampelopsis)
    A. arborea (L.) Koehne—pepper-vine
    A. cordata Mich.—heartleaf ampelopsis
Parthenocissus Planch. (woodbine)
    P. quinquefolia (L.) Planch.—Virginia creeper

Linden Family (Tiliaceae)
Tilia L. (basswood)
    T. americana L.—American basswood
    T. caroliniana Mill.—Carolina basswood
    T. heterophylla Vent.—white basswood

Mallow Family (Malvaceae)
Hibiscus L. (hibiscus)
    H. syriacus L.—rose-of-Sharon

Tea Family (Theaceae)
Stewartia L. (stewartia)
    S. malacodendron L.—silky camelia
    S. ovata (Cav.) Weatherby—mountain camelia

St. Johnswort Family (Hypericaceae)
Hypericum L. (St. Johnswort)
    H. densiflorum Pursh—dense-flowered St. Johnswort
    H. prolificum L.—shrubby St. Johnswort

Mezereum Family (Thymelaeaceae)

Dirca L. (leatherwood)
    D. palustris L.—leatherwood

Oleaster Family (Elaeagnaceae)
Elaeagnus L. (elaeagnus)
    E. umbellata Thunb.—autumn-olive

Ginseng Family (Araliaceae)
Hedera L. (ivy)
    H. helix L.—English ivy
Aralia L. (aralia)
    A. spinosa L.—devils-walkingstick

Tupelo-Gum Family (Nyssaceae)
Nyssa L. (tupelo)
    N. sylvatica Marsh.—black tupelo

Dogwood Family (Cornaceae)
  Cornus L. (dogwood)
    C. alternifolia L.—alternate-leaf dogwood
    C. amomum Mill.—silky dogwood
    C. florida L.—flowering dogwood

White-alder Family (Clethraceae)
  Clethra L. (clethra)
    C. acuminata Michx.—mountain pepperbush

Heath Family (Ericaceae)
  Chimaphila Pursh (prince's-pine)
    C. maculata (L.) Pursh—pipsissewa
  Rhododendron L. (rhododendron, azalea)
    R. arborescens (Pursh) Torr.—smooth azalea
    R. calendulaceum (Michx.) Torr.—flame azalea
    R. catawbiense Michx.—Catawba rhododendron
    R. maximum L.—rosebay rhododendron
    R. minus Michx.—Piedmont rhododendron
    R. nudiflorum (L.) Torr.—pink azalea
    R. vaseyi (Gray)—pinkshell azalea
    R. viscosum (L.) Torr.—swamp azalea
  Menziesia Smith (menziesia)
    M. pilosa (Michx.) Juss.—minnie-bush
  Leiophyllum (Pers.) Hedwig f. (sand-myrtle)
    L. buxifolium (Bergius) Ell.—sand-myrtle
    L. buxifolium var. *prostratum* (Loudon) Gray—Allegany sand-myrtle
  Kalmia L. (kalmia)
    K. angustifolia L.—lambkill
    K. latifolia L.—mountain-laurel
  Pieris D. Don (fetter-bush)
    P. floribunda (Pursh) B. & H.—fetter-bush
  Lyonia Nutt. (lyonia)
    L. ligustrina (L.) DC.—male-berry
  Leucothoe D. Don (sweetbells)
    L. editorum Fern. & Shub.—dog-hobble
    L. racemosa (L.) Gray—sweetbells
    L. recurva (Buckl.) Gray—mountain sweetbells
  Oxydendrum DC. (sourwood)
    O. arboreum (L.) DC.—sourwood
  Epigaea L. (ground-laurel)
    E. repens L.—trailing-arbutus
  Gaultheria L. (wintergreen)
    G. procumbens L.—aromatic wintergreen

Gaylussacia HBK. (huckleberry)
    G. baccata (Wang.) K. Koch—black huckleberry
    G. dumosa (Andrz.) T. & G.—dwarf huckleberry
    G. ursina (M. A. Curtis) T. & G. ex Gray—buckberry
Vaccinium L. (blueberry)
    V. arboreum Marsh.—sparkleberry
    V. atrococcum (Gray) Porter—black highbush blueberry
    V. constablaei Gray—highbush blueberry
    V. corymbosum L.—northern highbush blueberry
    V. erythrocarpum Michx.—bearberry
    V. hirsutum Buckl.—hairy blueberry
    V. macrocarpon Ait.—cranberry
    V. stamineum L.—deerberry
    V. vacillans Torr.—lowbush blueberry

Ebony Family (Ebenaceae)
Diospyros L. (persimmon)
    D. virginiana L.—persimmon

Sweetleaf Family (Symplocaceae)
Symplocos Jacq. (sweetleaf)
    S. tinctoria (L.) L'Her.—sweetleaf

Storax Family (Styracaceae)
Halesia Ellis ex. L. (silverbell)
    H. carolina L.—Carolina silverbell

Olive Family (Oleaceae)
Fraxinus L. (ash)
    F. americana L.—white ash
    F. americana var. *biltmoreana* (Beadle) J. Wright ex Fern.—Biltmore ash
    F. pennsylvanica Marsh.—green ash
    F. profunda (Bush) Bush—pumpkin ash
Chionanthus L. (fringetree)
    C. virginicus L.—fringetree
Ligustrum L. (privet)
    L. sinense Lour.—Chinese privet
Syringa L. (lilac)
    S. vulgaris L.—common lilac

Dogbane Family (Apocynaceae)
Vinca L. (periwinkle)
    V. minor L.—periwinkle

Vervain Family (Verbenaceae)
   Callicarpa L. (beautyberry)
      C. americana L.—beautyberry

Nightshade Family (Solanaceae)
   Solanum L. (nightshade)
      S. dulcamara L.—bitter nightshade

Figwort Family (Scrophulariaceae)
   Paulownia Sieb. & Zucc. (paulownia)
      P. tomentosa (Thunb.) Sieb. & Zucc. ex Steud.—royal paulownia

Bignonia Family (Bignoniaceae)
   Campsis Lour. (trumpet vine)
      C. radicans (L.) Seemann—trumpet creeper
   Catalpa Scop. (catalpa)
      C. bignonioides Walt.—southern catalpa
      C. speciosa Warder ex Engelm.—northern catalpa

Madder Family (Rubiaceae)
   Cephalanthus L. (buttonbush)
      C. occidentalis L.—buttonbush
   Mitchella L. (partridgeberry)
      M. repens L.—partridgeberry

Honeysuckle Family (Caprifoliaceae)
   Diervilla Mill. (bush-honeysuckle)
      D. lonicera Mill.—northern bush-honeysuckle
      D. sessilifolia Buckl.—southern bush-honeysuckle
   Lonicera L. (honeysuckle)
      L. bella Zabel—belle honeysuckle
      L. canadensis Marsh.—fly honeysuckle
      L. dioica L.—mountain honeysuckle
      L. japonica Thunb.—Japanese honeysuckle
      L. sempervirens L.—coral honeysuckle
   Symphoricarpos Duhamel (snowberry)
      S. orbiculatus Moench.—coralberry
   Viburnum L. (viburnum)
      V. acerifolium L.—mapleleaf viburnum
      V. alnifolium Marsh.—hobblebush
      V. cassinoides L.—withe-rod
      V. dentatum L.—southern arrowwood
      V. dentatum var. *lucidum* Aiton—northern arrowwood

V. nudum L.—possumhaw
V. prunifolium L.—blackhaw
V. rufidulum Raf.—rusty blackhaw
Sambucus L. (elder)
    S. canadensis L.—American elder
    S. pubens Michx.—red-berried elder

Glossary

Achene. A small, dry, seedlike fruit.

Aerial rootlets (vine). Small roots produced on a stem aboveground; often used for climbing, as in poison-ivy.

Alternate. Arranged singly along a stem; one leaf, bud, or leaf scar at a node. Compare *opposite, whorled.*

Apex. The upper end or tip; the part opposite the base.

Appressed. Lying against or pressed closely to another surface; opposite of *divergent.*

Armed. Bearing thorns or prickles.

Aromatic. Having a pleasant odor.

Awn. A slender, bristlelike structure, usually at the tip of a leaf or bud scale.

Axil. The upper angle between a leaf (or a leaf scar) and the stem.

Axillary (buds, flowers, fruit). Borne in the upper angle between a leaf and the stem; also called *lateral.* Compare *terminal.*

Berry. A fleshy fruit containing more than one seed.

Biennial. Requiring two growing seasons to reach maturity.

Blade. The broad part of a leaf.

Bloom (twig, leaf, fruit). A thin, powdery coating of wax, easily rubbed off.

Blunt. Having margins forming an obtuse angle (more than ninety degrees).

Bristle. A coarse, stiff hair.

Bristly. Covered with coarse, stiff hairs.

Bud. A young and undeveloped leaf, flower, or stem, either protected by one or more scales (scaly bud) or lacking this protection (naked bud).

Bud scale. A small, dry, modified leaf or part of a leaf covering a bud.

Bundle scars. Small dots or lines within the leaf scar; they are the severed ends of vascular bundles left by a separated or fallen leaf.

Capsule. A dry fruit, containing several to many seeds, opening at maturity.

Catkin. A compact, cylindrical cluster of unisexual flowers, each flower in the axil of a bract.

Chambered pith. Pith divided crosswise into small compartments separated by partitions or membranes; as used in this book, pith whose compartments are either hollow or filled with looser tissue. If the twig is cut lengthwise, such pith appears ladderlike.

Ciliate. Fringed with hairs (cilia) along the margin.

Collateral (buds). Placed side by side.

Compound leaf. A leaf whose blade is divided into two or more smaller parts called leaflets, each usually having the general appearance of a leaf. Compare *Simple leaf.*

Conical. Cone shaped.

Creeping. Running along at or near the surface of the ground and rooting.

Crescent. The shape of a new moon.

Deciduous. Falling off. Used for woody plants whose leaves fall in autumn, and also for plant parts, such as stipules and outer bud scales, that fall off soon after they appear. Compare *evergreen, persistent.*

Divergent (buds). Extending out; leaning away from the stem; opposite of *appressed.*

Double-toothed. Said of leaves with teeth of two sizes, each larger tooth bearing one or two smaller teeth.

Drupe. A fleshy fruit with a pit or stone, such as a cherry or plum.

Ellipsoid (fruit, buds). Shaped like an American football.

Elliptical (leaf). Having rounded sides and pointed ends like the outline of an American football.

Entire (leaf). Smooth margined; not toothed or lobed, but may be ciliate.

Erect. In a vertical position.

Evergreen. Leaves remaining on the plant in winter, keeping their green color. Compare *deciduous.*

Fascicle. A small bundle or close cluster; for example, the needles of pine.

Fruit. The mature, fully developed ovary of a flower, containing one or more seeds.

Genus. A group of closely related species, such as birches or hickories; plural, *genera.*

Glabrous. Without hairs.

Glands. Minute secreting structures, sometimes associated with hairs.

Glandular. Furnished with glands or gland-tipped hairs.

Glaucous (leaf, twig, fruit). Covered with a white or bluish bloom, easily rubbed off, as in plums.

Half-evergreen. Plants retaining green or bronzy leaves in mild winters and protected locations.

Heart shaped (leaf, leaf base). Having the shape of a valentine heart; also refers to a leaf base with two rounded lobes.

Hollow (pith). Not solid or chambered; pith in which the soft tissue is lacking but the space is present; in such cases the pith is said to be hollow.

Keel. A sharp ridge or rib, resembling the keel of a boat, on some bud scales.

Lanceolate (leaf). Shaped like a lance; long and narrow but broadest at the base.

Lateral. Said of buds and branches that appear along the sides of a twig. Compare *terminal.*

Leaf bud. A bud containing leaves and stem but no flowers.

Leaflet. A leaflike division of a compound leaf.

Leaf scar. The scar left on a stem when a leaf falls.

Legume. A plant of the pea family; the one- to many-seeded fruit of a plant of that family.

Lenticel. A small corky area, usually appearing as a dot, on the bark of young stems of many woody plants, serving as a pore allowing the exchange of gases between the air and the living tissue in the stem; in some species, as in the cherries, the lenticels lengthen horizontally as the growing stem expands.

Linear. Long and narrow, with the sides parallel or nearly so.

Lobed (leaf). Said of leaves whose blades are divided into segments by shallow to deep notches, the segments too large to be called teeth.

Long pointed (leaf). Having the apex gradually tapering to a slender point; acuminate.

Midrib. The central rib or vein of a leaf or a leaflet.

Mucro. A short, small, abrupt tip.

Naked. Not covered; bud lacking bud scales, the rudimentary leaves exposed, or a catkin not enclosed in a bud.

Needle. A long, slender, more or less needle-shaped leaf.

Node. The place on a stem where a leaf, flower, or bud is attached.

Notched. Marked by a V-shaped indentation.

Oblong (leaf). Having a length more than twice the width and the sides parallel or nearly so.

Obovate (leaf). Reversely ovate; egg shaped in outline, the narrower end at the base.

Obovoid (fruit). Like an egg-shaped solid, the narrower end at the base.

Once compound. A compound leaf having a single set of undivided leaflets.

Opposite (leaves, buds, leaf scars). Borne two at a node, one on each side of the stem; denoted on older growth by similarly arranged twigs. Compare *alternate, whorled.*

Ovate (leaf). Egg shaped in outline, the broader end at the base.

Ovoid (bud, fruit). Like an egg shaped solid, the broader end at the base.

Palmate. Radiating from a common center, such as leaflets attached to the end of a petiole, as in buckeye, and not arranged in rows along an axis.

Palmately veined (leaf). Having several equally prominent veins branching out from or near the base of the blade.

Partition. A transverse plate in the *pith.*

Persistent (leaves, fruit, bud scales). Remaining on the plant beyond the period when such parts commonly fall.

Petiole. The leafstalk; in this book also used loosely to include the central axis (technically, the rachis) of pinnately compound leaves, as in ash and locust.

Pinnately compound (leaf). Having leaflets arranged along the sides of a petiole.

Pinnately veined (leaf). Having secondary veins arising from the midrib.

Pistillate. Bearing female flowers producing seeds and fruits but not pollen. Compare *staminate.*

Pith. The soft, spongy innermost tissue in a stem.

Pome. A fruit resembling an apple.

Prickle. A sharp-pointed outgrowth from the bark, not connected with the vascular system of the plant, as in blackberry. Compare *spine* and *thorn.*

Prostrate. Lying flat on the ground.

Raceme. A simple flower cluster with stalked flowers borne along a central stem, the lowest flowers opening first, as in black cherry.

Resin. A plant secretion, frequently aromatic, that is insoluble in water but soluble in ether or alcohol.

Revolute (leaf). Having the margin rolled under.

Rounded. Forming a full, sweeping arc.

Samara. A winged fruit, as in maple and ash.

Sap. As used in this book, the fluid that flows from a freshly cut petiole or twig.

Scale. One of the small, modified leaves covering a bud; one of the short, pointed, overlapping leaves of some conifers; any small, thin, or flat structure.

Scrambling. Said of climbing plants that do not twine or produce tendrils or aerial rootlets, such as rose.

Scurf. Minute scales on the surface of a leaf, bud, or twig.

Seed. The ripened ovule; the structure for producing a new plant.

Short pointed (leaf). Ending sharply in an acute angle (less than ninety degrees).

Shrub. A woody plant smaller than a tree, typically with several stems or trunks arising from a single base, and less than five meters (about fifteen feet) tall.

Silky. Having long, fine, appressed, glossy hairs.

Simple leaf. A leaf with a single blade, not compound or made up of leaflets. Compare *compound leaf.*

Single toothed (leaf). Bearing teeth all of one size.

Sinus (leaf). The space between two lobes.

Solid (pith). Having pith uniformly dense, not chambered or hollow.

Solitary. Borne singly, not in pairs or clusters.

Species. A particular kind of plant within a given genus, such as bitternut hickory, *Carya cordiformis* (hickory, or *Carya,* being the genus); plural, species.

Spine. A thorn. Compare *thorn* and *prickle.*

Spongy (pith). Porous, suggesting a sponge.

Sprawling. Spreading out irregularly.

Spur (twigs). A short, stubby branch, usually bearing leaves, crowded leaf scars, or both.

Stalked bud. A bud definitely narrowed at the base below the lowest scales, as in mountain alder, or below rudimentary leaves, as in witch-hazel.

Staminate. Bearing male flowers; producing pollen but not seeds and fruits. Compare *pistillate.*

Stem. The trunk, branch, or twig of a plant; not the stalk of a leaf, which is the petiole.

Stipules. Small appendages, variable in shape, borne in pairs at the base of leaf-petioles in some woody plants.

Stipule scars. Small scars left by deciduous stipules.

Straight across (leaf). Shaped as if cut off at the base.

Striate. Marked with fine, usually parallel lines or ridges.

Superposed (buds). One or more extra buds that appear above the true axillary buds; usually flower buds.

Tendril. A part of a stem or leaf modified to serve as a holdfast organ by coiling around a support, as in grape.

Terminal (buds, flowers, fruit). Borne at the tip of a stem or branch. Compare *lateral, axillary.*

Thorn. A stiff, woody, sharp-pointed structure; as used in this book, a modified organ—stem, leaf, or stipule.

Trailing. Creeping along the ground.

Translucent. Partly transparent; permitting the passage of light.

Tree. A woody plant larger than a shrub, typically with one main stem or trunk, and more than five meters (about fifteen feet) tall.

Twice compound (leaf). A compound leaf in which the leaflets are subdivided into more leaflets.

Twig. A small branch, usually consisting of several years' growth. See also *young twig.*

Twining (vines). Twisting or coiling around a support.

Two-ranked (leaves, leaf scars). In two vertical rows.

Unequal, unsymmetrical (leaf). Slanting or having unequal sides; especially describes the base of leaves where one side extends farther down on the petiole than the other.

Valvate (bud). Having scales of the bud meeting at the edges without overlapping.

Variety. A strain of plants that stand out by themselves but are not sufficiently distinct to be a separate species; for example, *Sassafras albidum* var. *molle.*

Wedge shaped (leaf). In the form of a wedge; broad above and tapering by straight sides to a narrow base.

Whorled (leaves, buds, leaf scars). Borne three or more at a node. Compare *alternate, opposite.*

Woody plant. A plant with stems that contain lignin (wood).

Woolly. Matted with dense, short, soft hairs.

Young twig. As used in this book, the growth of the current season (bearing leaves) or the following winter; a twig one year old or less.

References

Engler, A., and K. Prantl. *Die Naturliche Pflanzenfamilien.* 20 vols. Leipzig, 1897–1915; 2d. ed., 1924– (incomplete).

Gupton, Oscar W., and Fred C. Swope. *Trees and Shrubs of Virginia.* Charlottesville, VA, 1981.

Little, Elbert L., Jr. *Checklist of United States Trees (Native and Naturalized).* Agriculture Handbook No. 541. Washington, DC., 1979.

Muenscher, W. C. *Keys to Woody Plants.* Ithaca, NY, 1922, 1975.

Petrides, George A. *A Field Guide to Trees and Shrubs.* Boston, MA, 1958, 1972.

Preston, Richard J., and Valerie G. Wright. *Identification of Southeastern Trees in Winter.* Raleigh, NC, 1978.

Radford, Albert E., Harry E. Ahles, and C. Ritchie Bell. *Manual of the Vascular Flora of the Carolinas.* Chapel Hill, NC, 1964, 1979.

Shanks, R. E., and A. J. Sharp. *Summer Keys to Tennessee Trees.* Knoxville, TN, 1950, 1972.

Stupka, Arthur. *Trees, Shrubs, and Woody Vines of Great Smoky Mountains National Park.* Knoxville, TN, 1964, 1985.

Wharton, Mary E., and Roger W. Barbour. *Trees and Shrubs of Kentucky.* Lexington, KY, 1973.

Index

Note: First number refers to text, second number to illustration. Common names are indexed only to text.